Towards the Use of Natural Compounds for Crop Protection and Food Safety

Towards the Use of Natural Compounds for Crop Protection and Food Safety

Editors

Lisa Pilkington
Siew-Young Quek

MDPI • Basel • Beijing • Wuhan • Barcelona • Belgrade • Manchester • Tokyo • Cluj • Tianjin

Editors
Lisa Pilkington
University of Auckland
Zealand

Siew-Young Quek
University of Auckland
Zealand

Editorial Office
MDPI
St. Alban-Anlage 66
4052 Basel, Switzerland

This is a reprint of articles from the Special Issue published online in the open access journal *Foods* (ISSN 2304-8158) (available at: https://www.mdpi.com/journal/foods/special_issues/natural_compounds_food).

For citation purposes, cite each article independently as indicated on the article page online and as indicated below:

LastName, A.A.; LastName, B.B.; LastName, C.C. Article Title. *Journal Name* **Year**, *Volume Number*, Page Range.

ISBN 978-3-0365-3751-1 (Hbk)
ISBN 978-3-0365-3752-8 (PDF)

© 2022 by the authors. Articles in this book are Open Access and distributed under the Creative Commons Attribution (CC BY) license, which allows users to download, copy and build upon published articles, as long as the author and publisher are properly credited, which ensures maximum dissemination and a wider impact of our publications.

The book as a whole is distributed by MDPI under the terms and conditions of the Creative Commons license CC BY-NC-ND.

Contents

About the Editors vii

Lisa I. Pilkington
Towards the Use of Natural Compounds for Crop Protection and Food Safety
Reprinted from: *Foods* 2022, 11, 648, doi:10.3390/foods11050648 1

Manasweeta Angane, Simon Swift, Kang Huang, Christine A. Butts and Siew Young Quek
Essential Oils and Their Major Components: An Updated Review on Antimicrobial Activities, Mechanism of Action and Their Potential Application in the Food Industry
Reprinted from: *Foods* 2022, 11, 464, doi:10.3390/foods11030464 3

Robin Raveau, Joël Fontaine, Abir Soltani, Jouda Mediouni Ben Jemâa, Frédéric Laruelle and Anissa Lounès-Hadj Sahraoui
In Vitro Potential of Clary Sage and Coriander Essential Oils as Crop Protection and Post-Harvest Decay Control Products
Reprinted from: *Foods* 2022, 11, 312, doi:10.3390/foods11030312 29

Zhipeng Gao, Weiming Zhong, Ting Liu, Tianyu Zhao and Jiajing Guo
Global Proteomic Analysis of *Listeria monocytogenes*' Response to Linalool
Reprinted from: *Foods* 2021, 10, 2449, doi:10.3390/foods10102449 53

Reza Sadeghi, Fereshteh Heidari, Asgar Ebadollahi, Fatemeh Azarikia, Arsalan Jamshidnia and Franco Palla
High-Pressure Carbon Dioxide Use to Control Dried Apricot Pests, *Tribolium castaneum* and *Rhyzopertha dominica*, and Assessing the Qualitative Traits of Dried Pieces of Treated Apricot
Reprinted from: *Foods* 2021, 10, 1190, doi:10.3390/foods10061190 67

Concetta Maria Messina, Rosaria Arena, Giovanna Ficano, Mariano Randazzo, Maria Morghese, Laura La Barbera, Saloua Sadok and Andrea Santulli
Effect of Cold Smoking and Natural Antioxidants on Quality Traits, Safety and Shelf Life of Farmed Meagre (*Argyrosomus regius*) Fillets, as a Strategy to Diversify Aquaculture Products
Reprinted from: *Foods* 2021, 11, 2522, doi:10.3390/foods10112522 79

Youssef Rouphael, Giandomenico Corrado, Giuseppe Colla, Stefania De Pascale, Emilia Dell'Aversana, Luisa Ida D'Amelia, Giovanna Marta Fusco and Petronia Carillo
Biostimulation as a Means for Optimizing Fruit Phytochemical Content and Functional Quality of Tomato Landraces of the San Marzano Area
Reprinted from: *Foods* 2021, 10, 926, doi:10.3390/foods10050926 99

Alejandro Jiménez-Gómez, Ignacio García-Estévez, M. Teresa Escribano-Bailón, Paula García-Fraile and Raúl Rivas
Bacterial Fertilizers Based on *Rhizobium laguerreae* and *Bacillus halotolerans* Enhance *Cichorium endivia* L. Phenolic Compound and Mineral Contents and Plant Development
Reprinted from: *Foods* 2021, 10, 424, doi:10.3390/foods10020424 113

About the Editors

Lisa Pilkington, Senior Lecturer at the School of Chemical Sciences, University of Auckland, graduated with a PhD from the same university in 2015 before completing an MSc in Applied Statistics at the University of Oxford (UK) in 2016. Her current research interests include the development and application of analytical and statistical methods, tools and techniques to answer chemical questions, applying these principles to a range of different areas, spanning agricultural/food, biological and medicinal/forensic applications. She is the recipient of a range of fellowships, including the JSPS HOPE Fellowship (2022), BCNZ Ronald Kay Fellowship (2020), Rutherford Post-Doctoral Fellowship (2019), Lottery Health Research Fellowship (2017), Gavin and Anne Kellaway Medical Research Fellowship (2015) and New Zealand Federation of Graduate Women Fellowship (2015).

Siew-Young Quek, the Director of Food Science Programme at University of Auckland, graduated with a PhD degree from School of Chemical Engineering, University of Birmingham (UK), in 1999. Her current research interests include microencapsulation of bioactives for delivery into functional foods, extraction and characterisation of bioactives/functional ingredients from food and waste by-products, processing and food quality, and food product development. She has received numerous awards and recognition for her research, including Shanghai Outstanding Overseas Scholar (2014/15), gold medal at the Geneva International Exhibition of Invention, New Technology & Products (2005) and University Invention & Research Competition in Malaysia. She is a fellow of the New Zealand Institute of Food Science and Technology.

Editorial

Towards the Use of Natural Compounds for Crop Protection and Food Safety

Lisa I. Pilkington

School of Chemical Sciences, University of Auckland, Auckland 1010, New Zealand; lisa.pilkington@auckland.ac.nz; Tel.: +64-9-373-7599 (ext. 86776)

Citation: Pilkington, L.I. Towards the Use of Natural Compounds for Crop Protection and Food Safety. *Foods* **2022**, *11*, 648. https://doi.org/10.3390/foods11050648

Received: 16 February 2022
Accepted: 17 February 2022
Published: 23 February 2022

Publisher's Note: MDPI stays neutral with regard to jurisdictional claims in published maps and institutional affiliations.

Copyright: © 2022 by the author. Licensee MDPI, Basel, Switzerland. This article is an open access article distributed under the terms and conditions of the Creative Commons Attribution (CC BY) license (https://creativecommons.org/licenses/by/4.0/).

The six research articles/communications and one review that comprise this Special Issue, which concerns studies towards natural compound use for crop protection and food safety purposes, highlight the most recent research and investigations into this exciting area.

With an ever-increasing global population, the demand for food, and on food-production, is massive. It is critical that we are able to meet this demand and mitigate the risks and factors that challenge our ability to do so, including pestilence to food crops and biological threats to food safety, before food reaches the consumer. As such, the advancement of measures to both protect crops and facilitate the surety of safe food products to end-users is a research area of great interest and growing development.

Essential oils are concentrated volatile extracts from plants and are, thus, a mixture of natural products. The application and use of EOs is seen as a method to increase the shelf-life of highly perishable foods by inhibiting pathogen proliferation. Due to the increasing interest in their use, Angane and co-workers [1] review recent research on EOs, focusing on the antibacterial activity of fruit-peel EOs, and the mechanism of action of EO components, as well as providing an overview of the recent contributions of EOs in food matrices. As stated in the review, research to date is extremely encouraging, and further research to devise strategies for EO application at an industrial scale is recommended.

As a specific example of essential oils and recent research into them, Raveau et al. [2] assess clary sage and coriander essential oils for their antifungal, herbicidal and insecticidal activities against notable plant pathogens and pests. It was found that these essential oils were able to inhibit the growth of a range of fungi, and that they exerted anti-germinative, herbicidal, repellent and fumigant effects. Furthermore, it was found that essential oil made from the aerial section of coriander exhibited the most significant antifungal and herbicidal effects. This work highlights that these essential oils have notable potential to act as crop protectants and control post-harvest decay.

Linalool is a vital component of many essential oils and has been particularly noted for its activity against *Listeria monocytogenes* (LM), one of the most serious foodborne pathogens that is responsible for the onset of listeriosis. In their work, Gao et al. [3] conduct an iTRAQ-based quantitative proteomic analysis to investigate the response of LM when exposed to linalool, in order to ascertain information about linalool's mode of action. GO and Kyoto Encyclopedia of Genes (KEGG) enrichment analysis, in conjunction with flow cytometry data, presented cell membranes, cell walls, nucleoids and ribosomes as putative targets of linalool against LM.

In addition to essential oils and their components, other natural chemicals and procedures are also being explored to prolong the shelf-life and traits of produced foods. An example of this includes the study of the use and effects of carbon dioxide treatment to control warehouse pests that commonly affect dried apricots, by Sadeghi et al. [4] In their research article, this group describes their investigation into the use of carbon dioxide gas at varying pressures to control two pest species: *Tribolium castaneum* (Herbst) and *Rhyzopertha dominica* (F.). In addition, the effect of CO_2 gas on the quality characteristics of dried apricots were assessed. Overall, it was found that CO_2 gas has the potential to protect dried apricots from *T. castaneum* and *R. dominica* during storage, without any significant impact on the product quality.

The storage and quality, this time of meagre fillets after being treated by cold-smoking, were explored by Messina and co-workers [5]. The smoked fillets were stored for 35 days at 4 °C, after which a range of analyses were performed to assess their sensory, biochemical, physical–chemical and microbiological properties. The results showed positive effects on the fillets' biochemical parameters and lipid peroxidation. Overall, this work highlights the potential to produce cold-smoked meagre as a value-added fish product.

In order to meet the rising demand for products, a range of methods are being explored to increase crop yield and production. Biostimulation is one such technique, and Rouphael et al. [6] explore the effect of plant biostimulation on fruits of the traditional tomato germplasm, which has been largely unexplored until now. They investigated how a tropical plant-derived biostimulant influenced the nutritional, functional, and compositional characteristics of tomato fruits, by profiling primary and secondary metabolites in the fruit. Biostimulation affected fruits from the different landraces differently, in many cases leading to improved yield and fruit quality, thus highlighting biostimulation as a promising method to optimise fruit yield and quality.

Jiménez-Gómez et al. [7] explore another potential method to increase crop production: the replacement of chemical fertilisers with biofertilisers (including plant-root-associated beneficial bacteria). In their research article, they describe their work, which assesses the use of *B. halotolerans* SCCPVE07 and *R. laguerreae* PEPV40 strains as efficient biofertilisers for escarole crops. It was shown that these two strains promoted plant development, and the escarole plants showed an increase in a range of minerals and constituents.

The innovative and exciting research included in this Special Issue highlights the interest and potential of this emerging area, addressing some of the most pressing global issues that we face, including sustainably feeding our ever-increasing population.

Funding: There is no funding to declare.

Conflicts of Interest: The authors declare no conflict of interest.

References

1. Angane, M.; Swift, S.; Huang, K.; Butts, C.A.; Quek, S.Y. Essential Oils and Their Major Components: An Updated Review on Antimicrobial Activities, Mechanism of Action and Their Potential Application in the Food Industry. *Foods* **2022**, *11*, 464. [CrossRef] [PubMed]
2. Raveau, R.; Fontaine, J.; Soltani, A.; Mediouni Ben Jemâa, J.; Laruelle, F.; Lounès-Hadj Sahraoui, A. In Vitro Potential of Clary Sage and Coriander Essential Oils as Crop Protection and Post-Harvest Decay Control Products. *Foods* **2022**, *11*, 312. [CrossRef] [PubMed]
3. Gao, Z.; Zhong, W.; Liu, T.; Zhao, T.; Guo, J. Global Proteomic Analysis of Listeria monocytogenes' Response to Linalool. *Foods* **2021**, *10*, 2449. [CrossRef] [PubMed]
4. Sadeghi, R.; Heidari, F.; Ebadollahi, A.; Azarikia, F.; Jamshidnia, A.; Palla, F. High-Pressure Carbon Dioxide Use to Control Dried Apricot Pests, Tribolium castaneum and Rhyzopertha dominica, and Assessing the Qualitative Traits of Dried Pieces of Treated Apricot. *Foods* **2021**, *10*, 1190. [CrossRef] [PubMed]
5. Messina, C.M.; Arena, R.; Ficano, G.; Randazzo, M.; Morghese, M.; La Barbera, L.; Sadok, S.; Santulli, A. Effect of Cold Smoking and Natural Antioxidants on Quality Traits, Safety and Shelf Life of Farmed Meagre (Argyrosomus regius) Fillets, as a Strategy to Diversify Aquaculture Products. *Foods* **2021**, *10*, 2522. [CrossRef] [PubMed]
6. Rouphael, Y.; Corrado, G.; Colla, G.; De Pascale, S.; Dell'Aversana, E.; D'Amelia, L.I.; Fusco, G.M.; Carillo, P. Biostimulation as a Means for Optimizing Fruit Phytochemical Content and Functional Quality of Tomato Landraces of the San Marzano Area. *Foods* **2021**, *10*, 926. [CrossRef] [PubMed]
7. Jiménez-Gómez, A.; García-Estévez, I.; Escribano-Bailón, M.T.; García-Fraile, P.; Rivas, R. Bacterial Fertilizers Based on Rhizobium laguerreae and Bacillus halotolerans Enhance Cichorium endivia L. Phenolic Compound and Mineral Contents and Plant Development. *Foods* **2021**, *10*, 424. [CrossRef] [PubMed]

Review

Essential Oils and Their Major Components: An Updated Review on Antimicrobial Activities, Mechanism of Action and Their Potential Application in the Food Industry

Manasweeta Angane [1,2,3], Simon Swift [2], Kang Huang [1], Christine A. Butts [3] and Siew Young Quek [1,4,*]

1. Food Science, School of Chemical Sciences, The University of Auckland, Auckland 1010, New Zealand; mang207@aucklanduni.ac.nz (M.A.); kang.huang@auckland.ac.nz (K.H.)
2. Faculty of Medical and Health Sciences, School of Medical Sciences, The University of Auckland, Auckland 1010, New Zealand; s.swift@auckland.ac.nz
3. The New Zealand Institute for Plant & Food Research Limited, Palmerston North 4442, New Zealand; chrissie.butts@plantandfood.co.nz
4. Riddet Institute, New Zealand Centre of Research Excellence for Food Research, Palmerston North 4474, New Zealand
* Correspondence: sy.quek@auckland.ac.nz; Tel.: +64-9-923-5852

Abstract: A novel alternative to synthetic preservatives is the use of natural products such as essential oil (EO) as a natural food-grade preservative. EOs are Generally Recognized as Safe (GRAS), so they could be considered an alternative way to increase the shelf-life of highly perishable food products by impeding the proliferation of food-borne pathogens. The mounting interest within the food industry and consumer preference for "natural" and "safe" products means that scientific evidence on plant-derived essential oils (EOs) needs to be examined in-depth, including the underlying mechanisms of action. Understanding the mechanism of action that individual components of EO exert on the cell is imperative to design strategies to eradicate food-borne pathogens. Results from published works showed that most EOs are more active against Gram-positive bacteria than Gram-negative bacteria due to the difference in the cell wall structure. In addition, the application of EOs at a commercial scale has been minimal, as their flavour and odour could be imparted to food. This review provides a comprehensive summary of the research carried out on EOs, emphasizing the antibacterial activity of fruit peel EOs, and the antibacterial mechanism of action of the individual components of EOs. A brief outline of recent contributions of EOs in the food matrix is highlighted. The findings from the literature have been encouraging, and further research is recommended to develop strategies for the application of EO at an industrial scale.

Keywords: essential oil; peel; antibacterial; antimicrobial; mechanism of action; preservation

1. Introduction

Antimicrobial agents used to kill or inhibit the growth of pathogenic or food spoilage bacteria can exist in natural or synthetic forms. The use of synthetic antimicrobial compounds as food preservatives has raised consumers' concerns, since they present numerous toxicological difficulties and may not be safe for human consumption [1]. Hence, over the last two decades, natural antimicrobial agents such as essential oils (EOs) have received renewed interest from the scientific community, owing to their unique physicochemical properties and diverse biological activities [2]. In the definition coined by Rios [3], EOs are aromatic, oil-like volatile substances present in plant materials such as fruits, bark, seeds, pulp, peel, root and whole plant. These substances form in the cytoplasm, and generally exist as tiny droplets sandwiched between the cells. In recent years, increasing awareness about the "green, safe and clean" environment and a growing appeal for "green consumerism" have prompted the production of foods free of synthetic preservatives [4,5].

EOs have been used for medicinal purposes and as therapeutic agents since ancient times [6]. Although food industries utilize EOs as a flavoring agent, their potential as a natural food grade preservative has not been fully explored. EOs present a valuable tool for food preservation due to their natural antimicrobial properties [7]. However, a detailed understanding regarding individual components of EOs, their antibacterial properties, mechanism of action and target organisms is required to support the implementation of EOs as food preservatives. Calo et al. [8] reported that EOs comprise numerous compounds such as aromatic hydrocarbons, terpene (monoterpenes and sesquiterpenes), terpenoids, esters, alcohols, acids, aldehydes and ketones, and their antibacterial activity is not solely contributed by any one compound. Recognizing the most potent antibacterial compounds from EOs is often tricky due to their chemistry complexity. To date, most studies have focused on studying the antimicrobial activity of EOs [5,8], with little discussion on the antibacterial activity of individual components in the EO or their mechanism of action. The antibacterial activity of EOs is not reliant on one specific mode of action; instead, EOs can attack several targets in a cell to inactivate the bacterium [7]. Evaluating EO's antibacterial properties and mechanism of action of their components may provide new insights into their applications in the food industry. This approach may reveal the concealed antibacterial properties of individual EO components, otherwise masked when EOs are studied as one single substance.

Several reviews [2,9,10] have outlined the antimicrobial activity of EOs extracted from various plant sources such as stem, bark, leaf, fruit, and seeds, but did not discuss the waste parts such as peel. The amount of waste produced by fruit processing industries is diverse [11]. Fruit peels generated by food industries are treated as agro-waste and are discarded in landfills, composted or fed to livestock [12]. Fruit waste produced in enormous quantities during commercial processing could present severe environmental threats [13]. Ayala-Zavala et al. [14] proposed using fruit by-products as an antimicrobial food additive, reporting that mandarins, papayas, pineapple, and mangoes accounted for 16.05%, 8.47%, 13.48% 11% of peel waste, respectively. On the other hand, fruit peel is a rich source of EOs and contains promising novel components of potential pharmacological, pharmaceutical and economic significance [13]. Moreover, fruit peel EOs are classified as GRAS (generally recognized as safe) and can be used to improve food safety due to their unique antimicrobial properties [15].

Studies on EOs extracted from various plant sources are well represented in the literature, and it is widely recognized that EOs possess a range of biological activities. For instance, EOs extracted from thyme [16], oregano, lavender [4,17], cinnamon, clove [18] and turmeric [19] have antibacterial, antifungal, algicidal, antioxidant, anticancer and anti-inflammatory activities. Chemical compositions and biological properties of plant EOs, in general, have been discussed in detail in reviews by Bakkali et al. [20], Burt [5] and Ju et al. [21]. A substantial amount of work has been carried out to evaluate the antimicrobial properties of EOs extracted from fruit peels; however, none of the reviews in the compiled data have exclusively discussed peel EOs. In light of these factors, this review aims to summarize the most significant findings of the antimicrobial properties of fruit peel EOs and their major components that contribute to microbial inactivation, with a focus on the mode of action of EO/EOs components. Finally, the application of various plant-derived EOs in the food industry is discussed, and future research directions and applications are presented.

2. Chemical Composition of Fruit Peel Essential Oils

Plants produce a variety of chemical compounds with antimicrobial properties. Some of these compounds are always present, while others are secreted in response to stress, such as infection, damage, predators, and weather variations. The chemical constituents in EOs are prone to variations depending on the time of harvest, cultivar, and the extraction method. Hydro distillation and steam distillation are frequently used to produce EOs at a commercial scale [5]. Identifying the most active compounds from EO can be a cumbersome process. Gas chromatography (GC), gas chromatography-mass spectrometry (GC-MS) [22–24], high-performance liquid chromatography (HPLC) [25–27] and liquid chromatography coupled to mass spectrometry (LC-MS) [28] are the most widely used methods to study the chemical composition of EOs. The primary chemical components of EOs are terpenes and polyphenols. Figure 1 shows the structural formula of some of the major components of EOs. These chemical compounds have been reported to have antimicrobial properties and their mechanisms of action are discussed later (Section 4).

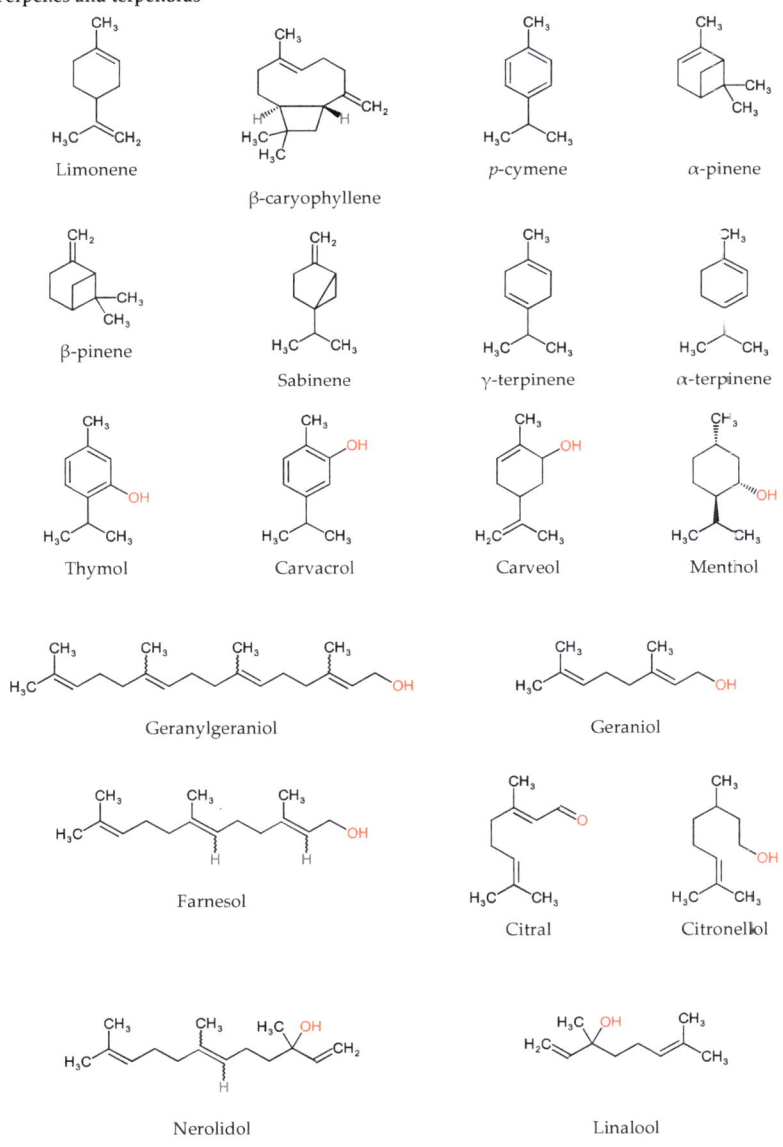

Figure 1. *Cont.*

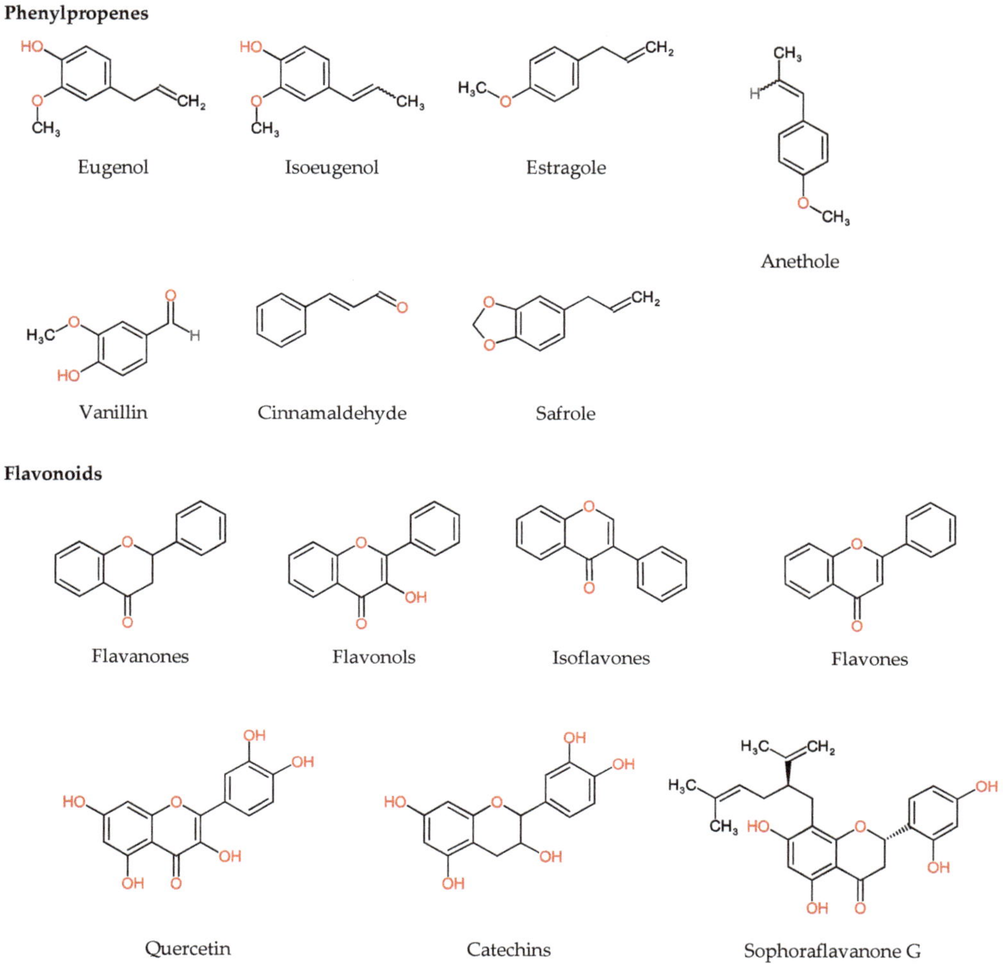

Phenylpropenes

Eugenol Isoeugenol Estragole Anethole

Vanillin Cinnamaldehyde Safrole

Flavonoids

Flavanones Flavonols Isoflavones Flavones

Quercetin Catechins Sophoraflavanone G

Figure 1. Chemical composition of essential oils (EOs).

Terpenes can be defined as a framework of numerous isoprene units (C_5H_8) merging to form a hydrocarbon molecule. They are derived from mevalonate and mevalonate-independent pathways [29]. Terpenes usually exist in EOs in the form of monoterpenes ($C_{10}H_{16}$) or sesquiterpenes ($C_{15}H_{24}$). However, other long-chain molecules such as diterpenes ($C_{20}H_{32}$), triterpenes ($C_{30}H_{48}$), tetraterpenes ($C_{40}H_{64}$) are found in EOs in minor quantities [30]. Examples of terpene compounds include β-caryophyllene, p-cymene, α-pinene, β-pinene, limonene, sabinene, γ-terpinene, α-terpinene, β-myrcene, cinnamyl alcohol, and δ-3-carene. Additionally present are terpenoids, identified as an oxygenated derivative of terpene compounds with an additional oxygen molecule, or their methyl group being moved or eliminated. Terpenoids are further categorized into esters, aldehydes, ketones, alcohols, ethers, and epoxides, with examples including menthol, geraniol, eugenol, thymol, carvacrol, geraniol, linalyl acetate, linalool, citronellal, citronellol and terpineol [7,31].

Polyphenols are secondary metabolites widely distributed in nature, usually derived from the phenylpropanoid pathway [32]. Polyphenols can be categorized into phenylpropenes and flavonoids, based on the number of phenol rings [33]. Phenylpropenes have derived their name from the six-carbon aromatic phenol group, and the three-carbon propene tail of cinnamic acid formed during the first step of phenylpropanoid biosynthesis [34]. Flavonoids are a group of phenolic compounds with a carbon framework (C_6-C_3-C_6). The basic skeletal structure of flavonoids comprises a 2-phenyl-benzo-γ-pyrone consisting of two benzene rings (ring A and ring B) cross-linked to a heterocyclic pyrone (ring C) [35]. Based on the degree of oxidation, flavonoids are further classified into flavones, flavonols, flavanones and others [36].

A detailed analysis of the EOs of orange peel identified an abundant amount of limonene, ranging between 73.9–97.6%, while other monoterpenic alcohols, namely geraniol, linalool, nerol and α-terpineol, were present in minor quantities at concentrations of 2.1%, 4.1%, 1.5%, 2.4%, respectively [24]. This finding was in agreement with Ambrosio et al. [22] and Guo et al. [37], who reported similar compounds in orange peel EOs. However, some compounds such as cis-p-mentha and trans-p-mentha [22,37] were not reported previously [24]. These differences could be attributed to the different cultivars or growing conditions of the fruit analyzed in these studies. Moreover, a close resemblance was noted in the limonene content of grapefruit peel EO, which was present at a concentration of 93.3% [23], 91.5% [38] and 91.8% [39]. Other monoterpene compounds such as β-myrcene, α-pinene, sabinene, linalool and thujene were also reported [23,38,39]. In pummelo peel EO, limonene contributed up to 55.7% of the total EO composition, followed by β-pinene (14.7%), linalool (6.2%), β-citral (4.1%), germacrene-D (2.7%), α-pinene (2.3%), α-terpineol (2.0%), geraniol (1.6%), sabinene (1.3%) [39]. Tao et al. [40] reported similar compounds but at a much lower concentration, ranging from 0.08% to 0.63%. The difference in the extraction method, such as using a rotary evaporator at 40 °C [38], could have contributed to the significant loss of highly volatile compounds from the EO. Furthermore, Hosni et al. [41] and Hou et al. [42] found limonene to be the main component in mandarin peel EO, but other secondary compounds such as lauric acid, 1-methyl-1,4-cyclohexadiene, methyl linoleate, myristic acid, palmitic acid and β-myrcene were reported only by Hou et al. [42]. More recent evidence [43] highlights that out of 158 compounds found in feijoa peel EO, 89 compounds identified were novel; these compounds include esters, sesquiterpenes, monoterpenes, aromatic hydrocarbons, alcohols, aldehydes, ketones, hydrocarbon, acids and ethers.

Limonene is the predominant component in the EOs of orange [22,24,37,41], grapefruit [23,39], mandarin and pummelo [39,40] peels, and is thought to contribute to most of the antimicrobial activity of the fruit peels reviewed above [44]. However, Ambrosio et al. [22] argued that limonene is present in different concentrations in different fruit peels; thus, the antimicrobial activity of EOs cannot be ascribed solely to limonene. Additionally, studies have reported low antimicrobial activity of limonene when the pure compound was tested [45]. Hence, in citrus fruits, other minor compounds such as α-pinene, sabinene, linalool, β-citral, and germacrene-D could contribute to the antimicrobial activity.

3. Antimicrobial Properties of Fruit Peel Essential Oils

The antimicrobial activity of EOs can be seen as the inhibition of cell growth or by cell-killing. However, it is not easy to differentiate between these modes of action. The antimicrobial efficacy of EOs is dependent on their chemical composition, environmental conditions and the structures of the target bacteria (either Gram-positive or Gram-negative bacteria) [46]. Numerous in vitro techniques [47], such as the determination of minimum inhibitory concentration (MIC) and minimum bactericidal concentration (MBC) by broth macro dilution/microdilution or agar disk/well diffusion are applied to determine the efficacy of an antimicrobial compound. Agar disk/well diffusion and broth macro dilution/microdilution are widely used methods in clinical microbiology laboratories [48] and have recently been recognized as useful tools to determine the antimicrobial activity of EOs [49,50].

Many studies have illustrated the antimicrobial effect of fruit peel EOs against drug-resistant, pathogenic and food spoilage bacterial strains. Some studies have found that EOs extracted from the fruit peels of banana [13], pomegranate [1] and citrus fruits such as sweet orange, grapefruit, lime, sweet lemon, mandarin, tangerine and pummelo [22,40,51–53] exhibited inhibitory activity against Gram-positive and Gram-negative bacteria. These studies indicate that fruit peels are a potentially valuable anti-microbial resource [42]. A wide range of foodborne pathogens could be inhibited by fruit peel EOs, including *Escherichia coli, Enterobacter cloacae, Klebsiella pneumoniae, Pseudomonas aeruginosa, Salmonella enterica* serovar Typhimurium, *Salmonella enteritidis, Bacillus subtilis, Bacillus cereus, Streptococcus faecalis, Listeria monocytogenes, Proteus vulgaris, Staphylococcus aureus* and others (Table 1). An overview of the antimicrobial activity of various fruit peel EOs and detection methods over the last 15 years is presented in Table 1.

3.1. Citrus Essential Oils

Abd-Elwahab et al. [51] reported the efficacy of EOs extracted from citrus peels, i.e., orange, lime, mandarin, and grapefruit, as having moderate to high antibacterial activity against *S. aureus, B. subtilis, E. faecalis, E. coli, Neisseria gonorrhoeae* and *P. aeruginosa*. Among those citrus EOs, lime peel EO was the most effective at inhibiting all six strains of pathogenic bacteria. The presence of coumarine and tetrazene in lemon peel [13] and citral, limonene and linalool in other citrus peel EO [51] may have accounted for their antimicrobial activity against these bacteria. On the contrary, Javed et al. [15] reported that amongst all tested citrus peel EOs (mandarin, tangerine, sweet orange, lime, grapefruit) mandarin peel EO possessed the highest antimicrobial activity. The inhibition zone for *Salmonella enterica* serovar Typhi, *E. coli, Streptococcus sp.* and *P. fluorescence* ranged from 20 to 30 mm for 10 µL and 9–16 mm for 5µL treatments of mandarin peel EO. The differing concentrations of the citrus peel EOs between the studies might explain these contradictory results.

3.2. Orange Essential Oils

Over the past decade, several studies [24,37,53–55] have examined the antibacterial properties of sweet orange (*Citrus sinensis*) EO. A broad-spectrum antibacterial activity was observed against a range of foodborne pathogens, confirming its potential to be a natural antimicrobial agent for food preservation. In a study conducted by Guo et al. [37], the antimicrobial activity of cold-pressed and light phase EO extracted from orange peel was compared using *E. coli, S. aureus,* and *B. subtilis*. It was reported that light phase EO showed a better antimicrobial activity compared to the cold-pressed EO. The higher antimicrobial activity can be attributed to a higher quantity of carvone and limonene in the light phase EO. Nwachukwu et al. [56] tested the efficacy of orange peel EO extracted using water and ethanol (hot and cold) against *E. coli, S. aureus,* and *Bacillus* sp. It was noted that hot ethanol extracted EO was more effective than the water extracted EO at inhibiting the three bacteria strains. Hot ethanol might have facilitated the better release of volatile compounds present in orange peel EO. These findings are similar to those of Ali et al. [55], Bendaha et al. [52], and Kirbaslar et al. [57], who reported similar antimicrobial activity of orange (*Citrus aurantim*) peel EO against *L. monocytogenes, S. aureus, E. coli, E. faecalis, B. cereus, K. pneumoniae* and *P. aeruginosa*. One of the significant drawbacks of these studies [15,37,51–53,55–57] was that they fail to consider the MIC and MBC values, thus providing no foundation for EO application in food. However, Geraci et al. [24] and Tao et al. [54] had reported the MIC values of orange peel EO, and as anticipated Gram-positive (*B. cereus, B. subtilis, S. aureus*) bacteria were reported to be more susceptible to the orange peel EO compared to the Gram-negative (*E. coli* and *P. aeruginosa*) bacteria.

3.3. Grapefruit Essential Oils

The antimicrobial activity of grapefruit (*Citrus paradisi*) peel EO against *B. subtilis, E. coli, S. aureus, S. enterica* serovar Typhimurium and *P. aeruginosa* was reported by Deng et al. [23]. It was noted that Gram positive *B. subtilis* was the most sensitive amongst all strains investigated, while Gram negative *P. aeruginosa* was the least sensitive organism. This antibacterial activity may be attributed to the presence of abundant limonene in the grapefruit peel EO [44]. Similarly, pummelo (*Citrus grandis*) peel EO showed good inhibitory activity against Gram-positive bacteria (MIC- 9.38 µL/mL) and moderate activity against Gram-negative bacteria (MIC- 37.50 µL/mL) [40]. Terpene alcohols such as linalool are known for their inhibitory activity against Gram-negative bacteria [58]. Although a substantial amount of linalool was found in the pummelo peel EO, it did not inhibit *E. coli* [40]. This microorganism was only susceptible to pure linalool, but not to EO with linalool as one of the components in a mixture of compounds [59]. The use of EO instead of linalool alone might have contributed towards a higher MIC value of pummelo peel EO against *E. coli*.

3.4. Essential Oils from Other Fruit Peels

Several researchers have examined the antibacterial activity of various other fruit peels such as tamarillo [60], bergamot (*Citrus bergamia*) [57,61], sweet lemon (*C. limetta*) [53,62], *C. deliciosa* [63], kumquat (*C. japonica*) [64] and feijoa (*Acca sellowiana*) [65]. Surprisingly, Diep et al. [60] and Mandalari et al. [61] revealed that the tamarillo and bergamot peel flavonoids, respectively, exhibited strong antibacterial activity against Gram-negative bacteria such as *E. coli*, *Pseudomonas putida*, *S. enterica* serovar Typhimurium and *P. aeruginosa*, while Gram-positive bacteria (*B. subtilis*, *L. innocua*, *S. aureus*) were resistant. Similarly, El-Hawary et al. [63] found that *C. deliciosa* EO extracted from its leaves and peel was more effective against Gram-negative bacteria than the Gram-positive bacteria. In contrast, the inhibitions zones for bergamot peel EO (11mm to 16mm) with no clear distinction between Gram-positive and Gram-negative bacteria [57], and sweet lemon EO, demonstrated good antibacterial activity against both Gram-positive and Gram-negative bacteria with inhibition zones measuring between 10 to 35 mm [62].

Due to the difference in their cell wall structure [34], Gram-positive bacteria are more susceptible to EOs than Gram-negative bacteria [23,40,54,66]. However, published data have shown no clear differentiation between Gram-positive and Gram-negative bacteria [60,63]. The reason for this contradictory result is discussed in Section 4. It is somewhat surprising that many studies have assessed the antimicrobial activity by using only the agar disk/well diffusion method [15,22,39,50–53,55,56,60,63,65,67–72]. Agar disk/well diffusion is a quick typing tool used to determine the sensitivity of the bacterial strain. However, this quick typing tool cannot differentiate between bacteriostatic and bactericidal effects. The agar disk/well diffusion is a preliminary method that is not suitable to determine MIC or MBC, since it becomes quite challenging to measure the amount of EO diffused in the medium. Moreover, the hydrophobic nature of EO might pose an added challenge with regard to its ability to diffuse through the media, potentially resulting in uneven distribution. On the other hand, though tedious and time-consuming, broth macro dilution or microdilution methods allow quantifying the exact antimicrobial agent concentration that is effective against the pathogen and visibly distinguishes between bacteriostatic and bactericidal effects [49]. Most of the studies reviewed so far tend to overlook the importance of the broth dilution method for the determination of MIC and MBC of EOs, which is vital for determining the exact concentration required to kill bacteria, a prerequisite for assessing their potential application in food preservation.

Table 1. Overview of antimicrobial activities of fruit peel essential oils (EOs) and extracts.

Source of Peel EO	Target Organism	Method Used	Solvent Used	Test Concentration	Remarks	References
Tamarillo	E. coli, P. aeruginosa, S. pyogenes, S. aureus	Disk diffusion	MilliQ, n-hexane, ethanol, methanol	115 µL of 100 mg/mL on 13 mm disk	E. coli was most sensitive to aqueous extract from the peel (inhibition zone of 24 mm), P. aeruginosa was most sensitive to methanol extract.	[60]
Grapefruit	B. subtilis, E. coli, S. aureus, S. enterica serovar Typhimurium, P. aeruginosa	Disk diffusion, MIC determination	-	20 µL of 100, 50, 25, 12.5, 6.25, 3.125, 1.56, 0.78, 0.39 and 0.195 mg/mL of EO placed on each disk	B. subtilis represented a maximum inhibitory zone of 35.59 mm and MIC value of 0.78 µL/mL. P. aeruginosa was least sensitive representing an inhibition zone of 8.57 mm and MIC value of 25.0 µL/mL	[23]
Sweet orange, Lemon, Banana	P. aeruginosa, K. pneumoniae, Serratia marcescens, E. coli, P. vulgaris, S. enterica serovar Typhi, S. aureus, E. faecalis, L. monocytogenes, Aeromonas hydrophila, Streptococcus pyogenes, Lactobacillus casei	Agar well diffusion, MIC determination	Distilled water, Methanol, Ethanol, Ethyl acetate	5 mg/mL	K. pneumoniae was most susceptible to lemon peel extract (inhibition zone and MIC, 35 mm, and 130 µg/mL, respectively).	[13]
Kumquat	E. coli, S. enterica serovar Typhimurium, S. aureus, P. aeruginosa	Disk diffusion and MIC determination by broth microdilution method	Methanol 80%, Ethanol 70%, Acetone, Ethyl acetate, n-Hexane, Chloroform	From 10 mg/mL, 25 µL of extract was placed on each disk.	For all extracts, E. coli was most resistant (inhibition zone 11.3 mm and MIC of 679 µg/mL) while S. aureus was the most susceptible (inhibition zone 16.7 mm and MIC of 496 µg/mL) strain.	[64]
Sweet orange	enterotoxigenic E. coli, Lactobacillus sp	Disk diffusion and MIC determination	EO solutions prepared at 90% (v/v), using acetone	7 µL of EO solution placed on each disk	EO showed higher antimicrobial activity against ETEC, no activity shown against beneficial Lactobacillus sp.	[22]
Lemon	B. subtilis, E. coli, S. enterica serovar Typhimurium, S. aureus	Disk diffusion method	-	0.1 mL of EO solution placed on each disk	Ripened lemon peel EO was more effective against all four strains than the unripe lemon peel EO.	[71]
Sweet orange	Bacillus sp., E. coli, S. aureus	Agar well diffusion method	Hot ethanol, Cold ethanol, Hot aqueous, Cold aqueous	50 and 100 µL of each extract placed on disk	Hot ethanolic extract (100 µL) most effective, showing inhibition zone of 16, 15 and 16 mm against Bacillus sp., E. coli and S. aureus, respectively.	[56]
Feijoa	E. coli, S. aureus	Agar well diffusion method	Water and Methanol extracts	100 µL of each extract placed on disk	Methanol extract was more effective (Inhibition zone for E. coli and S. aureus was 14.7 and 26.5 mm, respectively).	[65]
Sweet orange	S. aureus, B. subtilis, E. coli	MIC determination by tube dilution method	Light phase and cold-pressed EO	-	The MIC of light phase EO for S. aureus, B. subtilis and E. coli was 3.13, 1.56 and 0.78 µL/mL, respectively.	[37]
Sweet orange	S. aureus, L. monocytogenes, P. aeruginosa	MIC determination by agar dilution method	EO and hexane extracts	100 to 2.5 mg/mL	EO was effective against L. monocytogenes (MIC value of 15 mg/mL) but less active against S. aureus and P. aeruginosa. Hexane extract at 10 mg/mL concentration was most effective.	[24]
Sweet orange, Lime, Mandarin, Grapefruit	B. subtilis, S. aureus, E. faecalis, E. coli, P. aeruginosa, N. gonorrhoeae	Disk diffusion and MIC determination by agar dilution method	-	10 µL of EO solution placed on each disk	Lime peel was most effective. MIC of 14 and 11 µL/mL was recorded for S. aureus and E. coli, respectively.	[51]

Table 1. Cont.

Source of Peel EO	Target Organism	Method Used	Solvent Used	Test Concentration	Remarks	References
Sour orange, Sweet orange, Grapefruit, Lemon	S. aureus, E. coli, E. faecalis, B. cereus	Agar well diffusion method	Aqueous extract	50 µL of 100 mg/mL of extract was dispensed in each well	The inhibition zones for S. aureus, E. faecalis, B. cereus and E. coli ranged from 10 to 18 mm, 9 to 17 mm, 11 to 18 mm, and 14 to 21 mm, respectively.	[55]
Bitter orange	L. monocytogenes, S. aureus, E. coli DH5α, Citrobacter freundii	Disk diffusion	Hexane extract	-	S. aureus was moderately sensitive to bitter orange extract (inhibition zone of 10mm). The extract did not inhibit Gram-negative organisms.	[52]
Grapefruit, Pummelo	E. coli, P. aeruginosa, S. enterica subsp., S. aureus, E. faecalis	Cold-pressed and water-distilled extracted EO		100 µL of 10 and 20 mg/mL of EO was suspended in each well	20mg/mL of pummelo peel EO presented antimicrobial activity against Gram negative Salmonella enterica subsp. followed by E. faecalis > E. coli > S. aureus > P. aeruginosa.	[39]
Sweet orange, Sweet lemon, Lemon	S. enterica serovar Typhimurium, E. coli	Disk diffusion method	Hexane extract	-	The inhibitory zone for S. enterica serovar Typhimurium and E. coli ranged from 4 mm to 10 mm.	[53]
Pomegranate	S. aureus, E. aerogenes, S. enterica serovar Typhimurium and K. pneumoniae	Agar well diffusion method	Methanol, Ethanol (100, 70, 50, 30%), Water	10 µL of extract: water (1:6) was dispensed in each well	S. aureus was the most sensitive strain, followed by E. aerogenes, S. enterica serovar Typhimurium, K. pneumoniae. The inhibition zone for S. aureus ranged from 24.5 to 20.3 mm.	[69]
Banana	S. aureus, S. pyogenes, Enterobacter aerogenes, K. pneumoniae, E. coli, Moraxella catarrhalis	Agar well diffusion	Aqueous extract	-	S. aureus showed an inhibition zone of 30 mm, but E. coli was resistant to the extract.	[67]
C. deliciosa	S. aureus, Micrococcus luteus, E. coli, P. vulgaris	Agar diffusion method	-	15 µL of EO was dispensed on the agar surface	The inhibition zone for all tested organisms ranged from 8 mm to 30 mm.	[63]
Lemon, Sweet lemon	S. aureus, S. epidermidis, S. agalactiae, E. faecalis, Streptococcus pneumoniae, S. pyogenes, E. coli, E. aerogenes, K. pneumoniae, Proteus sp., S. enterica serovar Typhimurium, Acinetobacter sp., Moraxella catarrhalis, P. aeruginosa	Agar well diffusion method	Aqueous extract	20 µL of extract was dispensed in each well	The effect of lemon and sweet lemon peel on microbial isolates was not significantly different. The inhibition zone for lemon and sweet lemon ranged from 20–30 mm and 10–35 mm, respectively.	[62]
Grapefruit	B. cereus, S. faecalis, E. coli, K. pneumoniae, Pseudo-coccus sp., S. enterica serovar Typhimurium, Shigella flexneri, S. aureus	Agar well diffusion method	Methanol, Ethanol	100 µL of 8, 40 and 80 µg/mL concentrations of EO solutions were dispensed in each well	Methanol extract was more effective against all tested strains. B. cereus was the most sensitive bacteria (inhibition zone from 30.33 to 32.67 mm), while E. faecalis was the most resistant one (inhibition zone from 6.0 to 12.0 mm)	[72]
Pomegranate	B. subtilis, S. aureus, E. coli, K. pneumoniae	Microdilution method	Methanolic and aqueous extracts	0.097–12.5 mg/mL	The MIC value for the tested strains ranged from 0.2 to 0.78 mg/mL.	[73]
Pummelo	S. aureus, B. subtilis, E. coli	Disk diffusion and MIC determination by broth microdilution method	-	10 µL of 50% (v/v) EO was placed on each disk. MIC concentration ranged from 1.17 to 750 µL/mL (v/v).	The inhibition zones for B. subtilis, S. aureus and E. coli were 17.08, 11.25 and 8.27 mm, respectively. The MIC values for B. subtilis, S. aureus and E. coli were 9.38 and 37.50 µL/mL, respectively.	[40]

Table 1. Cont.

Source of Peel EO	Target Organism	Method Used	Solvent Used	Test Concentration	Remarks	References
Pomegranate	16 strains of *Salmonella sp.*	Disk diffusion and MIC determination	Ethanol	20 µL of 100, 200 and 500 µg/mL concentration of EO solution was placed on each disk. MIC concentration ranged from 3.9 to 2000 µg/mL	The inhibition zone and the MIC values for *Salmonella sp.* ranged from 13.3 to 18.8 mm and 62.5 to 1000 µg/mL, respectively.	[74]
Lemon	*P. aeruginosa, S. enterica* serovar Typhimurium, and *Micrococcus aureus*	Agar well diffusion method and MIC determination	Methanol, Ethanol, Acetone	Dilutions from crude extract were prepared as follows: 1:20, 1:40, 1:60, 1:80, 1:100	All concentrations of lemon peel extracts effectively inhibited all the three strains tested.	[75]
Mandarin, Tangerine, Sweet orange, Lime, Grapefruit	*E. coli, S. enterica* serovar Typhi, *K. pneumoniae, E. cloacae, P. fluorescence, Proteus myxofaciens, S. epidermidis, Streptococcus sp.*	Disk diffusion method	-	From 500 µg/mL of stock solution 5 and 10 µL of EO was placed on each disk.	*S. enterica* serovar Typhi and *P. myxofaciens* were susceptible to all citrus EO tested.	[15]
Grapefruit	*S. aureus, E. faecalis, S. epidermidis, E. coli, S. enterica* serovar Typhimurium, *S. marcescens* and *P. vulgaris*	Disk diffusion method	-	20 µL of extract was dispensed in each well	*S. enterica* serovar Typhimurium was the most resistant (15 mm) strain followed by *E. faecalis* (16 mm), *S. epidermis* (17 mm), *S. marcescens* (19 mm), *P. vulgaris* (21 mm) and *S. aureus* (53 mm).	[38]
Pomegranate	*E. coli, Pseudomonas fluorescens, S. enterica* serovar Typhimurium, *S. aureus, B. cereus*	MIC determination by tube dilution method	Water	Final concentration of 0.01, 0.05, 0.1% was prepared in saline	*S. aureus* and *B. cereus* got inhibited at a concentration of 0.01%, *P. fluorescens* at 0.1%, *E. coli* and *S. enterica* serovar Typhimurium were not inhibited.	[76]
Pomegranate	*L. monocytogenes, S. aureus, B. subtilis, E. coli, P. aeruginosa, K. pneumoniae, Yersinia enterocolitica*	Agar well diffusion and MIC determination by agar dilution method	Methanolic (80%) and water extracts	800 µg/100 µL of extract was suspended in each well. MIC concentration ranged from 0 to 4 mg/mL	The inhibition zone for methanolic extract ranged from 13–20 mm. MIC determination showed that *Y. enterocolitica* was the most sensitive strain representing MIC of 0.25 mg/mL.	[77]
Sour lime	*B. subtilis, B. cereus, S. aureus, E. coli, E. aerogenes S. enterica* serovar Typhimurium	Disk diffusion method	-	-	*B. subtilis, B. cereus, S. aureus, S. enterica* serovar Typhimurium, *E. coli* and *E. aerogenes* showed inhibition zones of 22, 19.8, 18, 17, 16 and 10.5 mm, respectively.	[68]
Sweet orange	*S. aureus, B. subtilis, E. coli*	Disk diffusion and MIC determination by broth microdilution method	-	10 µL of 50% (v/v) EO was placed on each disk. MIC concentration ranged from 1.17 to 750 µL/mL, (v/v).	The inhibition zones for *S. aureus, B. subtilis* and *E. coli* were 23.37, 18.89 and 17.21 mm, respectively. The MIC values for *S. aureus, B. subtilis* and *E. coli* were 4.66, 9.33 and 18.75 µL/mL, respectively.	[54]
Lemon, Grapefruit, Bitter orange, Sweet orange, Mandarin, Bergamot	*S. aureus, B. cereus, Mycobacterium smegmatis, L. monocylogenes, M. luteus, E. coli, K. pneumoniae, P. aeruginosa, P. vulgaris*	Disk diffusion method	-	20 µL of EO solution was placed on each disk	Lemon peel EO exhibited better antimicrobial activity towards all bacteria with inhibition zone ranging from 10 to 16 mm.	[57]
Bergamot	*E. coli, P. putida, S. enterica, L. innocua, B. subtilis, S. aureus, Lactococcus lactis*	MIC determined using Bioscreen C	Ethanol (70, 100%)	200–1000 µg/mL	The MIC values for *E. coli, S. enterica, P. putida* were 200, 400, 500 µg/mL, respectively. Gram-positive bacteria showed no effect.	[61]

4. Effect of Chemical Components of Essential Oils on Food Spoilage and Pathogenic Microbes

In the literature, various modes of antimicrobial activity of EOs against a range of bacteria have been discussed [5,7,78,79]. However, before investigating the effect of fruit peel EO on microbes, we should have a closer look at the cell-wall structure of Gram-negative and Gram-positive bacteria (Figure 2).

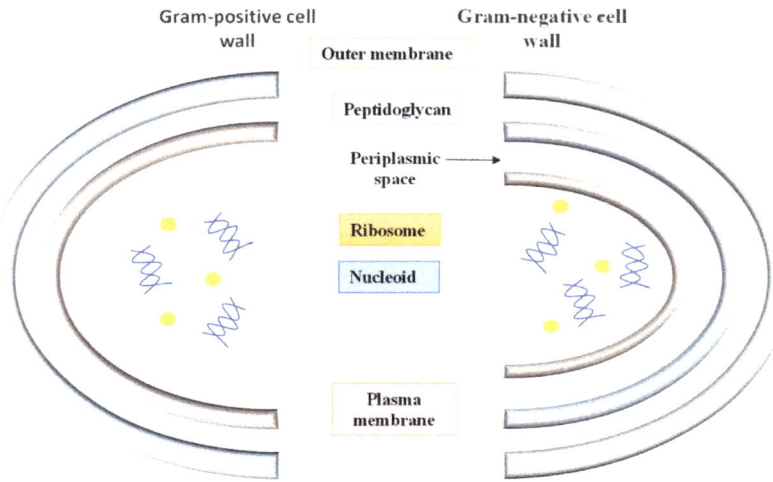

Figure 2. Schematic representation of Gram-positive and Gram-negative bacterial cell wall.

The hypothesis that Gram-positive bacteria are more susceptible to the effect of hydrophobic compounds such as EOs was first proposed by Plesiat et al. [80] followed by Nazzaro et al. [34], Chouhan et al. [79] and Raut et al. [81]. The difference between the susceptibility is attributable to the fact that Gram-positive bacteria have a thick layer of peptidoglycan linked to other hydrophobic molecules such as proteins and teichoic acid. This hydrophobic layer surrounding the Gram-positive bacterial cell may facilitate easy entry of hydrophobic molecules. On the other hand, Gram-negative bacteria have a more complex cell envelope comprising an outer membrane linked to the inner peptidoglycan layer via lipoproteins. The outer membrane contains proteins and lipopolysaccharides (lipid A), making it more resistant to the hydrophobic molecules in EO [82].

Other researchers investigating the antimicrobial activity of EOs showed no notable difference between the MIC values of Gram-positive and Gram-negative bacteria [13,39,60,61,64]. Although it has been hypothesized that the outer membrane is almost impermeable to the hydrophobic compounds, Plesiat et al. [80] argued that some hydrophobic compounds might cross the outer membrane via porin channels. Similarly, Van de Vel et al. [58] believe that some EO molecules are more active against Gram-positive bacteria, while others are active against Gram-negative bacteria, but the mechanisms remain unknown. Most studies on the antimicrobial activity of EOs have used *E. coli* and *S. aureus* as model microorganisms to represent Gram-negative and Gram-positive bacteria, respectively [65,83,84]. This could lead to a generalization of results, as not all Gram-negative and Gram-positive bacteria would follow a similar trend as observed in *E. coli* and *S. aureus*. Furthermore, the mode of action of EO depends on its chemical profile and the ratio of its active components [85]. The possible mechanisms wherein EOs interfere with bacterial proliferation may involve the following: (1) the disintegration of the bacterial outer membrane or phospholipid bilayer, (2) alteration of the fatty acid composition, (3) increase in membrane fluidity resulting in leakage of potassium ions and protons; (4) interference with glucose uptake, and (5) inhibition of enzyme activity or cell lysis (Figure 3) [5,86].

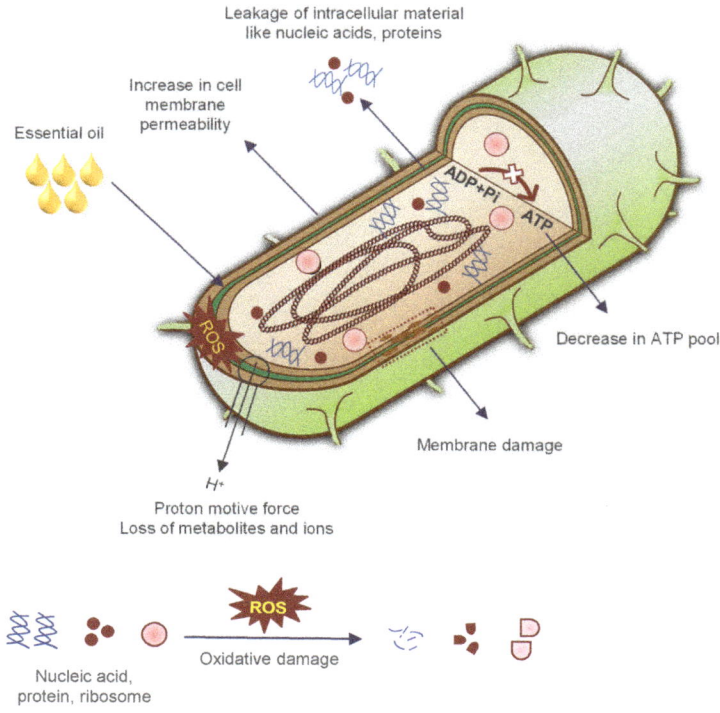

Figure 3. Antibacterial mechanism of essential oils (EOs).

In general, fruit peel EOs may comprise more than a hundred compounds [43]. Major compounds can contribute around 85–95% of the total EO composition, while other minor compounds can be present in trace amounts. While these compounds may have specific antimicrobial effects, Cho et al.'s [86] review draws attention to the synergistic and additive effect minor compounds might have in combination with the other components. Terpenes and terpenoids are primary components of essential oil followed by polyphenols [32]. Here, we discuss the antimicrobial activity and mode of action of EOs and their components on the bacterial cell.

4.1. Terpenes and Terpenoids

Terpenes and terpenoids constitute a significant class of compounds in EOs known to have antimicrobial activity. The potential antimicrobial activity of thymol and carvacrol has been extensively discussed in previous reviews [7,34,79]; hence we exclude them from our discussion to focus on other EO compounds. Thymol and carvacrol are the major components of thyme and oregano oil, respectively, and are structurally analogous differing in the location of hydroxyl groups on the phenol ring [7].

It is well recognized that terpenes can disrupt the lipid assembly of the bacterial cell wall, leading to disintegration of the cell membrane, denaturation of cell proteins, leakage of cytoplasmic material, which ultimately causes cell lysis and cell death [47]. Kim et al. [87] were amongst the first to show the antimicrobial potential of EO components including citral, limonene, perillaldehyde, geraniol, linalool, α-terpineol, carvacrol, citronellal, eugenol, β-ionone and nerolidol against *E. coli*, *S. enterica* serovar Typhimurium, *L. monocytogenes* and *Vibrio vulnificus*. It was suggested that terpenes and terpenoids might interfere with oxidative phosphorylation or oxygen uptake in microbial cells, thereby inhibiting microbial growth [88]. Later, this hypothesis was supported by Zengin and Baysal's study [89], wherein terpene compounds such as linalool, α-terpineol and eucalyptol were reported to damage the cell membrane and alter the morphological structure of *S. aureus*, *S. enterica* serovar Typhimurium and *E. coli* O157:H7. The plausible explanation for this observation was that these terpene compounds interacted with the membrane proteins and phospholipids, leading to cellular respiratory chain inhibition, interruption in oxidative phosphorylation, disruption of nucleic acid synthesis, and loss of metabolites [90].

Two studies conducted by Togashi et al. [90,91] examined the effect of geranylgeraniol, geraniol, nerolidol, linalool and farnesol on *S. aureus*. All these terpene alcohols were reported to have antibacterial activity, with farnesol and nerolidol demonstrating the most potent antibacterial activity as determined by the broth dilution technique. They also explored the mechanism of these terpene alcohols on the bacterial cell membrane by measuring the leakage of K$^+$ ions from the bacterial cell, anticipating that distortion of the bacterial cell membrane leads to leakage of K$^+$ ions, thus indicating the presence of membrane disrupting compounds. In support of this, Akiyama et al. [92] reported the strong inhibitory effect of farnesol against *S. aureus*. Farnesol has also exhibited notable antibacterial activity against biofilms of *S. aureus* and *S. epidermidis* [93,94]. Akiyama et al. [92] attributed these inhibitory effects of farnesol to its hydrophobic nature, which accumulates in the cell membrane, thus disrupting the cell membrane as illustrated by scanning electron microscopy (SEM). Furthermore, an ester compound of geranyl acetate makes it a more potent antimicrobial compound than its parent moiety (geraniol), purportedly due to its hydrophobicity [95]. However, past studies [31,96] have demonstrated the antimicrobial activity and mechanism of geraniol, rather than geranyl acetate. For instance, geraniol was noted to inhibit *E. coli* and *S. aureus* [97], and multidrug-resistant *Enterobacter aerogenes* by acting as an efflux pump inhibitor [96,98]. Similar, to farnesol, it is thought that the antimicrobial potential of geraniol was due to its hydrophobic nature.

Han et al. [44] and Liu et al. [99] examined the antibacterial mechanism of limonene on *L. monocytogenes* and the antibacterial mechanism of linalool on *P. aeruginosa*, respectively. In their analysis, Han et al. [44] and Liu et al. [99] demonstrated that the compounds distorted the cell wall structure of bacteria and led to leakage of intracellular molecules such as nucleic acids and proteins, which also affected the functionality of the respiratory chain complexes and hampered the process of adenosine triphosphate (ATP) synthesis. Moreover, Gao et al. [100] elaborated the anti-listeria activities of linalool against its planktonic cells and biofilms using RNA-sequence analysis. Other articles have discussed the antimicrobial efficacy of limonene [101] and linalool [102] against various strains of microorganisms. The antimicrobial activity of limonene is due to the presence of alkenyl substituent and a double bond in the molecular structure that enhances its antimicrobial activity [95]. Other authors proposed that the cell membrane may be an important site for linalool to inactivate the cell [100]. The interaction causes thickening of the Gram-positive cell wall, eventually leading to cell disruption [103]. The S (+) enantiomer of linalool enables it to interact with the negatively charged outer membrane of the Gram-negative cell, thus facilitating the easy entry of the compound into the intracellular space, leading to disruption [104].

Dorman et al. [105] tested 14 EO compounds against 25 strains of bacteria and reported that monoterpenoid and sesquiterpene demonstrate potent antimicrobial activity against most strains tested. In the same way, Trombetta et al. highlighted the antimicrobial potential of monoterpenes (linalyl acetate, thymol and menthol) against *E. coli* and *S. aureus* [106]. The hydroxyl group present in the compound may have contributed to its antimicrobial activity. Guimaraes et al. [31] evaluated 33 terpene compounds commonly isolated from EOs for their antimicrobial efficacy, of which only 16 compounds were reported to possess antibacterial activity. Scanning electron microscopy results revealed that individual components of EOs such as geraniol, citronellol, carveol, and terpineol altered the cellular morphology and destroyed the cell membrane. This is supported by two previous studies where similar compounds were found to be potent [105,106]. Lopez-Romero et al. [107] conducted a similar study wherein the antibacterial effect and mechanism of action of essential oil components such as carveol, carvone, citronellol, and citronellal were evaluated against

E. coli and *S. aureus*. Citronellol was found to be the most effective, which led to a change in the cell membrane integrity, the surface charge followed by leakage of K$^+$ ions. In another study, two pentacyclic triterpenes, namely α-amyrin and ursolic acid, were also reported to have a disorganizing effect on the *E. coli* cell membrane [108]. Additionally, Garcia et al. [66] listed five monoterpene compounds (citronellal, citral, α-pinene, isopullegol and L-carvone) which possessed antifungal properties against three fungal strains and suggested their potential use in tropical fruit preservation. Other researchers [109,110] have investigated the antimicrobial potential of a bicyclic sesquiterpene, i.e., β-caryophyllene, against a range of microorganisms. However, they were unable to explain for the antibacterial mechanism with their study.

4.2. Polyphenols

Studies on polyphenols extracted from various fruit sources are well represented in the literature, and it is acknowledged that polyphenols possess a range of antimicrobial activities against pathogenic microbes. For example, the polyphenols in the skin extracts of Italian red grape, plum and elderberries demonstrated strong inhibitory properties against *S. aureus, B. cereus, E. coli, L. monocytogenes* while showing a growth-promoting effect on beneficial microbes such as *Lactobacillus rhamnosus, L paracasei* and *Lactobacillus plantarum* [111].

4.2.1. Phenylpropenes

Although phenylpropenes account for a smaller proportion of total volatiles than terpenes and terpenoids, they have been noted to have a significant contribution to the antimicrobial activity of EOs [112]. Phenylpropenes are not only found in some fruit varieties such as apple peel [113], lemon peel [114] and grapefruit peel [115], but are also found in a wide variety of spices and herbs such as clove, star anise, sweet basil and fennel [116].

The antimicrobial potential of eugenol has been extensively investigated [117–120]. Eugenol is thought to alter the permeability of the cell membrane, followed by leakage of intracellular ATP and macromolecules such as protein and nucleic acids, ultimately leading to cell death [119]. This theory was supported by Cui et al.'s [118] study wherein eugenol permeabilized the cell membrane leading to leakage of intracellular macromolecules and enzymes such as β-galactosidase, ATP and alkaline phosphatase (AKP). Furthermore, Qian et al. [117] noted that eugenol demonstrates cell membrane permeability properties and presents potent inhibition against the biofilm formation of *K. pneumoniae* cells. Likewise, Ashrafudoulla et al. [119] reported antibiofilm activity against *Vibrio parahaemolyticus* and cell membrane damaging properties, which led to leakage of cell contents. Research by Nazzaro et al. found that isoeugenol worked in a similar way to eugenol [34]. Hyldgaard et al. [121] explained that isoeugenol formed hydrogen bonds with the lipid headgroup, thus disturbing the lipid structure and destabilizing the membrane. This mechanism of action is known as a "non-disruptive detergent-like mechanism", and the free hydroxyl group and the molecule's hydrophobic nature were considered accountable for their antimicrobial activity [122]. However, Gharib et al. [112] argued that hydrophobicity might not be the only factor contributing to the molecule's antimicrobial activity, since in his study, eugenol and isoeugenol demonstrated a fluidizing effect on the bacterial cell wall. Furthermore, Auezova et al. [123] and Gharib et al. [112] examined the mechanism of allylic (eugenol and isoeugenol) and propenylic (estragole and anethole) phenylpropenes on the cell wall of *E. coli* and *Staphylococcus epidermidis*. They demonstrated the distinctive ability of estragole and anethole to penetrate the outer membrane of *E. coli*. The antimicrobial potency is conferred by the higher lipophilic nature of estragole and anethole (log P values of 3.5 and 3.4, respectively) in comparison to eugenol and isoeugenol (log P values of 2.5 and 3.0, respectively).

Cinnamaldehyde has also demonstrated anti-biofilm activities against *S. epidermidis* [124]. Other researchers have studied the antibacterial mechanism of cinnamaldehyde against *E. coli*, *S. aureus* [125] and *Aeromonas hydrophila* [126], reporting that it caused cell membrane distortion and leakage, in addition to condensation and polarization of the cytoplasmic content. The antibacterial activity of vanillin was studied against *Mycobacterium smegmatis*, and it was able to enhance the cell membrane permeability and alter cell membrane integrity [127].

4.2.2. Flavonoids

Flavonoids are polyphenolic compounds with a benzo-γ-pyrone group and are ubiquitously found in plant cells [36]. Few examples of flavonoids are flavanones, flavan-3,4-diols, chalcones, flavan-3-ols, flavonols, flavones, isoflavones, catechins, quercetin, anthocyanidins and proanthocyanidins [128]. Recent evidence suggests that flavonoids possess antibacterial activities against plant pathogens and human pathogens. Their antimicrobial mechanism is similar to traditional drugs [33], and hence could be of importance for use as natural antimicrobial agents.

A study on catechins showed that the compounds caused oxidative damage in *E. coli* and *B. subtilis* cells, thus altering cell membrane permeability and damaging the cell membrane [129]. Moreover, Cushnie et al. [130] also reported that catechins were responsible for potassium ion leakage in methicillin-resistant *S. aureus* (MRSA), which is the primary signal of membrane damage, and Tsuchiya et al. [131] reported that sophoraflavanone G significantly affected the membrane fluidity of the bacterial cells.

5. Application of Essential Oils in Food Products
Preservation

Traditional food preservation methods include chilling, frozen storage, drying, salting, smoking and fermentation [132]. However, consumers have questioned techniques such as fermentation, brining, and salting, due to the increasing demand for reduced-salt foods [133]. The meat industries utilize chemical preservatives such as nitrate salt, sulfites, chlorides to inhibit the growth of foodborne pathogens. These compounds have been associated with carcinogenic effects and other health complications [133]. Hence, the options available to substitute chemical preservatives with natural compounds have attracted increased interest in recent years. Lucera et al. [134], in her review, outlined some natural preservatives of animal origin, (lactoferrin, lysozyme); bacteriocin from microbes (natamycin, nisin); natural polymers (chitosan); organic acids (citric and propionic acid); EOs and extracts derived from plants. In this context, EOs are attracting considerable attention due to their application as a natural bio-preservative and inhibitor in food matrices or food products. At present, the investigations have focused primarily on EOs from herbs and spices. There is limited research on fruit peel EOs. So, the discussion is widened here to cover the food applications of all plant-derived EOs. Some publications have investigated the potential contributions EOs/extracts to extend the shelf-life and to inhibit the growth of pathogens in fresh-cut vegetable mixtures [135], lettuce, purslane [136], fruit juices [137], ready to eat meat [138], chicken nuggets [76] and breast [139], minced beef [140,141] and turkey [142]. A literature review [141] published in 2018 included 2473 publications since 1990 on the antimicrobial activity of EOs. Many of these publications investigated the application of EO's on food products, including 657 papers on fruits, 403 on vegetables, 415 on fish products, 410 on meat products, 216 on milk and dairy products, and 97 on bread and baked foods [143]. Other recent reviews have discussed the application of rosemary extract in meat [144], the synergistic effect of EO in seafood preservation [145], application of EO in active packaging [146] and as a food preservative [147]. The following section includes the recent history of EO by restricting the citations to the last 5–6 years to provide the readers with an update on EOs and their application in the food matrix (Table 2).

As consumers have gained greater awareness on issues related to health, processing and food additives, demand for natural and minimally processed food has soared. However, maintaining the freshness of fruits and fresh-cut vegetables for extended periods has been challenging. Spraying, dipping, coating, and impregnation are ways EOs can be applied to fruits and vegetables for maintaining shelf-life [134]. Some recent examples of these approaches are discussed here. He et al. [148] evaluated the effects of dipping cherry tomatoes in thyme EO nanoemulsion (TEON) against *E. coli* O157:H7 and the effect of

TEON in combination with ultrasound treatment. Their study showed that TEON alone could effectively inhibit the growth of *E. coli* O157:H7 on the surface of cherry tomatoes, and there was a substantial synergistic effect of the combined treatment. Kang et al. [149] found that freshly cut red mustard leaves, when washed with cinnamon leaf EO nanoemulsion, reduced the count of *E. coli*, *L. monocytogenes*, *S. enterica* serovar Typhimurium by more than one log. Another study conducted by the same author showed that washing with cinnamon leaf EO nanoemulsion improved physical detachment and inhibited both *L. monocytogenes* and *E. coli* O157:H7 on kale leaves [150]. Both studies did not show any adverse changes in the quality attributes of mustard [149] and kale leaves [150]. The lettuce leaves examined during 7-day storage periods showed a reduction in *E. coli* O157:H7 population when rinsed with a combination of carvacrol/eugenol and thymol/eugenol when compared to the control (water rinse). However, the treatments had adverse effects on the sensory analyses [151]. In contrast, a combination of Spanish origanum oil and Spanish marjoram oil successfully inhibited *L. monocytogenes* from a mixture of fresh-cut vegetables without showing any adverse sensory attributes [135]. A recent study elucidated that *Litsea cubeba* EO added to bitter gourd, cucumber, carrot and spinach juices at MIC concentration decreased the counts of *E. coli* O157:H7 by 99.1%, 99.92%, 99.94%, 99.96%, respectively [152]. Krogsgård Nielsen et al. [153] tested the inhibitory potential of isoeugenol and encapsulated isoeugenol against *L. monocytogenes*, *S. aureus*, *Leuconostoc mesenteroides*, *P. fluorescens* in carrot juice. Contrary to expectations, their study did not find a significant difference in the inhibitory activity of encapsulated and non-encapsulated isoeugenol.

Besides fruits and vegetables, much work on the antimicrobial potential of EO was studied in meat products especially beef and beef products [154–158]. Pistachio EO [155] and *Melaleuca alternifolia* (tea tree) EO [157] reduced the total viable and total *L. monocytogenes* counts in ground beef. The efficiency of 5% and 10% clove EO on the inactivation of *L. monocytogenes* in ground beef at refrigeration (8 °C), chilling (0 °C) and freezing (18 °C) temperatures was investigated by Khaleque et al. [158]. They observed that 10% clove EO was a lethal concentration to inactivate *L. monocytogenes* irrespective of temperature conditions, but 5% clove EO was ineffective at inactivating the pathogen [158]. Similarly, Yoo et al. [154] found that 0.5%, 1.0% and 1.5% clove EO did not significantly reduce the count of *E. coli* O157:H7 and *S. aureus* in beef jerkies. However, their study took an additional step and demonstrated that the combined effect of clove EO with encapsulated atmospheric pressure plasma had a bactericidal effect on both pathogens. Likewise, a study conducted by Lin et al. [156] pointed out the synergistic effect of chrysanthemum EO incorporated into chitosan nanofibers which inhibited *L. monocytogenes* in beef at a rate of 99.9%.

A triple combination of thyme/cinnamon/clove EO in the food matrix was first applied experimentally by Chaichi et al. [159]. The triple combination at FIC of 0.3, 0.39.0.43 had a bacteriostatic effect on *P. fluorescens* inoculated in chicken breast meat, while a triple combination at higher concentration (200 mg/kg) had an instant bactericidal effect. Thyme EO effectively inhibited *P. aeruginosa*, *E. coli* and *S. enterica* serovar Typhimurium in ground beef [160]. A recent study by Kazemeini et al. [161] prepared edible coatings of alginate containing *Trachyspermum ammi* EO (TAEO) as nanoemulsion to control the growth of *L. monocytogenes* in turkey fillets. The turkey fillets were coated with the emulsion and stored at 4 °C for 12 days. They observed the highest reduction of *L. monocytogenes* numbers in turkey fillets treated with 3% alginate containing 0.5% and 1% TAEO compared to non-coated samples. Other research articles have reported that EO nano emulsions effectively inhibited pathogens in rainbow trout fillet [162] and chicken breast fillets [163]. Apart from fruits, vegetables and meat products, the application of EO has also been evaluated on bakery [164,165] and dairy products [166].

Although several authors [152,157,165,166] have claimed successful testing for the application of EOs in different food systems, their approach has not escaped criticism. Santos et al. [167] emphasized the use of MBC concentration rather than MIC concentration in the food matrix to ensure a complete inhibition. These authors [167] questioned the usefulness of EOs in food systems because various factors such as environmental condition, age and cultivar of the plant, time harvested, extract composition and extraction method may impact the antimicrobial activity of the EO. All the above factors might challenge the rationale of applying EOs at a commercial level. Moreover, it is known that fat and protein present in food can solubilize or bind to phenolic compounds in EO, thus reducing its antimicrobial efficacy [157]. This view was supported by Khaleque et al. [158], who analyzed the effect of cinnamon EO at a higher concentration (2.5 and 5.0%) against *L. monocytogenes* in ground beef and found that cinnamon EO was unsuccessful in inactivating *L. monocytogenes* in ground beef. They also reported adverse organoleptic impacts upon using higher concentrations of EOs. In a study by Lages et al. [168], thyme EO combined with beet juice powder failed to give a desirable effect in reducing coagulase-positive *Staphylococcus* in meat sausage. It was recommended that combining half of the suggested dosage of chemical preservatives such as nitrites with EO could be feasible. Despite the question regarding the suitability of EO in minimally processed food products [167], only a few studies did not show effective inhibition by EOs of foodborne pathogens. In contrast, many studies have demonstrated the successful replacement of synthetic preservatives with EO in different food systems [165,166,169]. Since Santos et al. [167] did not use EOs in minimally processed food products, their assumptions need further validation. Their paper would have been more convincing if the authors had used food matrices to prove their hypothesis. There is evidence that EOs exhibit antimicrobial properties, therefore, their ability to be used as a natural preservative on an industrial scale needs further rigorous evaluation.

Table 2. Overview of recent studies on antimicrobial activity of different essential oils (EOs) in the food matrix.

Essential Oil	Pathogen	Food	Method Used	Concentration Applied	References
Thyme (*Thymus vulgaris*)	*E. coli* O157:H7	Cherry tomatoes	Dipping	0.0625, 0.125 mg/mL	[148]
Clove (*Syzygium aromaticum*)	*E. coli* O157:H7, *S. aureus*	Beef jerkies	Treated with EO and dried for 2 hrs	0.50%, 1.00%, 1.50%	[154]
Ajwain (*Trachyspermum ammi*)	*L. monocytogenes*	Turkey fillets	Coating	8, 4, 2 mg/mL	[161]
May chang (*Litsea cubeba*)	*E. coli* O157:H7	Bitter gourd, cucumber, carrot, and spinach juice	Inoculation	0.5, 0.25 mg/mL	[176]
Felon herb (*Artemisia persica* Boiss)	*L. monocytogenes*, *E. coli* O157:H7	Probiotic doogh	Addition of EO and mixing	75 ppm, 150 ppm	[164]
Rosemary (*Rosmarinus officinalis*), Lavender (*Lavandula*), Mint (*Mentha piperita*)	*Penicillium crustosum*	Bread	Exposing bread to a disk loaded with EO	125, 250, 500 µL/L	[165]
Thyme (Thy), Cinnamon (CN) (*Cinnamomum verum*), Clove (CV)	*P. fluorescens*	Chicken breast	Coated by dipping in EO emulsion for 5 min	Thy- 0.560 g/L, CN- 0.042, 0.170 g/L, CV- 0.078, 0.312 g/L	[154]
Ginger (*Zingiber officinale*), Clove, Thyme	*S. aureus*, *P. aeruginosa*, *E. coli*, *E. faecalis*, *P. fluorescens*, *C. albicans* and *Aspergillus parasiticus*	Fortified cheese	EO added and stirred	0.01%	[166]
Tea tree (*Melaleuca alternifolia*)	TVC, Psychrophilic, Coliform, Salmonella, Yeast, and mould count	Beef steaks	Addition of EO and mixing	0.1%, 0.5%	[155]
Cranberry extract (*Vaccinium macrocarpon*)	*Listeria* sp.	Chicken breast	Dipped in extract solution	4, 5 mg/mL	[170]
Thyme	Thermotolerant coliforms and *Escherichia coli*	Hamburger	Addition of EO and mixing	0.1 g/100 g of thyme EO 1 g/100 g of encapsulated thyme EO	[169]
Cinnamon leaf EO nanoemulsion	*L. monocytogenes*, *E. coli* O157:H7	Kale leaves	Washing	50 ppm	[149]
Thymol, Eugenol, Carvacrol	*E. coli* O157:H7	Lettuce leaves	Rinsing	0.63 mg/mL	[151]
Chrysanthemum (*Chrysanthemum indicum*)	*L. monocytogenes*	Beef	Packed into membrane (Chitosan nanofiber loaded with EO)	1.5%	[156]
Pistachio (*Pistacia vera*)	Total viable count (TVC)	Ground beef	EO added to meat and stomached for 1 min	1.5% (v/w)	[157]
Black cumin (*Bunium persicum*)	*E. coli* O157:H7	Rainbow trout fillet	Coated by dipping in nanoemulsion for 15 min	0.5%	[163]
Cranberry extract (*Vaccinium macrocarpon*)	Aerobic mesophilic count, *Brochothrix thermosphacta*, *P. putida*, *L. mesenteroides*, *L. monocytogenes*, *C. jejuni*	Pork meat slurry, hamburger, cooked ham	Mixed in meat	3.3%, 1.65%, 0.83%, 0.42%	[171]
Anise (*Pimpinella anisum*)	TVC, Psychrotropic count, Enterobacteriaceae, Lactic acid bacteria, *Pseudomonas* sp.	Minced beef	EO added using micropipette and massaged manually for 2 min	0.1%, 0.3%, 0.5% (v/w)	[172]
Coriander (*Coriandrum sativum*)	TVC, sulphite-reducing clostridia, *Salmonella* sp., *E. coli*, *L. monocytogenes*	Pork sausage	Mixed in sausage	0.000, 0.075, 0.100, 0.125, 0.150 µL/g	[173]
Cinnamon	TVC, Enterobacteriaceae	Italian pork sausage	Mixed in sausage	0.1%, 1.5% (v/v)	[174]
Ginger	Psychrophilic, Yeast and mould count	Chicken breast fillet	Coated by dipping in emulsion	3%, 6%	[162]
Cranberry extract	*E. coli*, *Salmonella enterica* serovar Enteritidis, *L. monocytogenes*, *S. aureus*	Minced pork		2.5 g/100 g	[175]
Thyme	*E. coli*, *S. enterica* serovar Typhimurium, *S. aureus*, and *P. aeruginosa*	Minced beef meat	EO added to meat and stomached for 5 min	0.001%, 0.05% 3% of EO in 10% DMSO (v/w)	[164]
Cinnamon EO (CEO) and grape seed extract (GSE)	TVC, Lactic acid bacteria, Psychrotropic count, Yeast, and mould count	Sausage	Mixed in sausage and packed in polyamide bags	CEO (0.02% and 0.04%) and GSE (0.08% and 0.16%)	[174]
Apple mint (*Mentha suaveolens*)	*E. coli*, *S. aureus*	Turkey sausage		2, 5, 10 mg/g	[177]
Isoeugenol	*L. monocytogenes*, *S. aureus*, *Leuconostoc mesenteroides*, *P. fluorescens*	Carrot juice	Inoculation	702, 1580 mg/mL	[151]
Thyme	*Salmonella enterica* serovar Enteritidis, *S. enterica* serovar Typhimurium, *S. Montevideo* and *S. Infantis*	Minced pork	Mixed in minced meat and vacuum packed	0.3%, 0.6%, 0.9%	[178]
Thyme	*L. monocytogenes*	Beef and pork sausage	Mixed and vacuum packed	100 ppm	[179]
Clove, Cinnamon	*L. monocytogenes*	Ground beef	Adding and mixing	Clove—5%, 10% Cinnamon—2.5%, 5%	[158]
Spanish origanum oil, Spanish marjoram oil and coriander oil	*L. monocytogenes*	Fresh cut vegetables	Immersing in EO solution	0.1%, 0.4%, 0.9%	[153]
Peppermint (*Mentha piperita*)	*Vibrio* spp.	Cheese	Applying on surface	5–15 µL/mL	[180]

6. Food Regulations on Applications of Essential Oils

The European Commission has documented a variety of EO compounds as approved flavour additives in different types of food products. In 2008, the European Commission released a list of approved compounds which is updated regularly. Some of the registered flavoring compounds that pose no risk to human health are limonene, linalool, β-caryophyllene, pinene, thymol, carvacrol, carvone, eugenol, isoeugenol, vanillin, citral, citronellal, cinnamaldehyde, menthol and lavandulol [181]. Moreover, the Food and Drug Administration (FDA) of the United States also recognizes these compounds as GRAS. Crude EOs such as mustard, oregano, clove, cinnamon, nutmeg, thyme, basil, rosemary and lavender are recognized as GRAS. The regulatory limits on acceptable daily intake on EO compounds and EOs are in place to govern their use in food products [7]. Despite the regulatory limits, EOs might cause allergic reactions and ingesting high doses of EOs or topical applications of EOs for a long period have been associated with severe health problems, such as oral toxicity and dermatitis [182]. Therefore, it is crucial to attain a fine balance between toxicity and effective dose in food products.

7. Conclusions and Future Prospects

Evidence from in vitro and in situ studies suggests that EOs possess good antibacterial activity against a wide range of foodborne pathogens. This review has evaluated studies on EOs that have the potential to act as natural preservatives in food products, due to their antioxidant and antimicrobial properties [183,184]. The potential of all plant-derived EOs, not just fruit peel EOs, has been evaluated for use as a preservative in foods. However, their application in food products have been restricted at an industrial scale as high doses are required to attain good antimicrobial activity, and the quantity, source and active composition profile of the EO to be used in food has not been optimized. In addition, components of the foods, such as fat [185], starch [186] and protein [187], may bind to the active compounds in EOs and reduce their efficacy. The volatile compounds in EOs may also produce undesirable chemical compounds by interacting with other food components such as proteins. To validate the use of EOs at an industrial level, the evaluation of these aspects is of paramount importance.

Firstly, high concentrations of EO in food have shown unappealing sensory attributes. However, this problem may be addressed by evaluating an effective synergistic/additive combination of EOs or a combination of EOs with other food preservation techniques such as temperature, irradiation, and pulse-electric field to reduce the required dosage of EO for the inhibition of pathogens. Another plausible solution for minimizing the interaction of EO compounds with food components such as fat, starch and proteins is by encapsulating the EO in an appropriate biodegradable material (e.g., chitosan), which might ensure controlled release without altering its biological activity. Secondly, a detailed understanding of how EOs work (the mechanism of action) will provide insights into the application of EO in the food industry to combat the proliferation of food-borne pathogens. To further study the mechanism of action, proteomic and transcriptomic analyses are needed to understand the pathways targeted by the EO compounds. The transition of in vitro experiments to in vivo trials to evaluate the efficacy of EOs has always posed an added challenge. Another future opportunity lies in the potential effects of EOs on immunity and gut health. Recent research reported that a combination of oregano extract with peppermint and thyme EO supported the growth of probiotic bacteria and positively affected the gut's microbial composition [188]. Further research regarding the role of EO on the gut microbiome would be worth exploring.

Author Contributions: Conceptualization, M.A., S.S. and S.Y.Q.; writing—original draft preparation, M.A.; writing—review and editing, M.A., S.S., K.H., C.A.B., S.Y.Q.; supervision, S.S., K.H., S.Y.Q.; project administration, S.Y.Q.; funding acquisition, S.S., S.Y.Q. All authors have read and agreed to the published version of the manuscript.

Funding: This research is partially funded by The University of Auckland (Press Account Number-9448-UOA-MANG207) and Food and Health Programme Seed Grant (4200-UOA-48422-A8AN).

Informed Consent Statement: Not applicable.

Acknowledgments: The authors would like to thank The University of Auckland for the Doctoral Scholarship awarded to the first author and Food and Health Programme Seed Grant.

Conflicts of Interest: The authors declare no conflict of interest.

References

1. Singh, B.; Singh, J.P.; Kaur, A.; Singh, N. Antimicrobial potential of pomegranate peel: A review. *Int. J. Food Sci. Technol.* **2019**, *54*, 959–965. [CrossRef]
2. Asbahani, A.E.; Miladi, K.; Badri, W.; Sala, M.; Addi, E.H.A.; Casabianca, H.; Mousadik, A.E.; Hartmann, D.; Jilale, A.; Renaud, F.N.R.; et al. Essential oils: From extraction to encapsulation. *Int. J. Pharm.* **2015**, *483*, 220–243. [CrossRef]
3. Rios, J.L. Essential Oils: What They Are and How the Terms Are Used and Defined. In *Essential Oils in Food Preservation, Flavor and Safety*; Preedy, V.R., Ed.; Academic Press: San Diego, CA, USA, 2016; Chapter 1; pp. 3–10.
4. Martucci, J.F.; Gende, L.B.; Neira, L.M.; Ruseckaite, R.A. Oregano and lavender essential oils as antioxidant and antimicrobial additives of biogenic gelatin films. *Ind. Crop. Prod.* **2015**, *71*, 205–213. [CrossRef]
5. Burt, S. Essential oils: Their antibacterial properties and potential applications in foods—A review. *Int. J. Food Microbiol.* **2004**, *94*, 223–253. [CrossRef] [PubMed]
6. Garzoli, S.; Petralito, S.; Ovidi, E.; Turchetti, G.; Laghezza Masci, V.; Tiezzi, A.; Trilli, J.; Cesa, S.; Casadei, M.A.; Giacomello, P.; et al. *Lavandula × intermedia* essential oil and hydrolate: Evaluation of chemical composition and antibacterial activity before and after formulation in nanoemulsion. *Ind. Crop. Prod.* **2020**, *145*, 112068. [CrossRef]
7. Hyldgaard, M.; Mygind, T.; Meyer, R.L. Essential oils in food preservation: Mode of action, synergies, and interactions with food matrix components. *Front. Microbiol.* **2012**, *3*, 12. [CrossRef]
8. Calo, J.R.; Crandall, P.G.; O'Bryan, C.A.; Ricke, S.C. Essential oils as antimicrobials in food systems—A review. *Food Control* **2015**, *54*, 111–119. [CrossRef]
9. Aleksic Sabo, V.; Knezevic, P. Antimicrobial activity of *Eucalyptus camaldulensis* Dehn. plant extracts and essential oils: A review. *Ind. Crop. Prod.* **2019**, *132*, 413–429. [CrossRef]
10. Papadochristopoulos, A.; Kerry, J.P.; Fegan, N.; Burgess, C.M.; Duffy, G. Natural anti-microbials for enhanced microbial safety and shelf-life of processed packaged meat. *Foods* **2021**, *10*, 1598. [CrossRef]
11. Joshi, V.; Kumar, A.; Kumar, V. Antimicrobial, antioxidant and phyto-chemicals from fruit and vegetable wastes: A review. *Int. J. Food Ferment. Technol.* **2012**, *2*, 123–136.
12. Chanda, S.; Barvaliya, Y.; Kaneria, M.; Rakholiya, K. Fruit and vegetable peels—Strong natural source of antimicrobics. In *Current Research, Technology and Education Topics in Apllied Microbiology and Microbial Biotechnology*; Mendez, V.A., Ed.; Formatex Research Center: Badajoz, Spain, 2010; Volume 1, pp. 444–450.
13. Saleem, M.; Saeed, M.T. Potential application of waste fruit peels (orange, yellow lemon and banana) as wide range natural antimicrobial agent. *J. King Saud Univ. Sci.* **2020**, *32*, 805–810. [CrossRef]
14. Ayala-Zavala, J.F.; Rosas-Domínguez, C.; Vega-Vega, V.; González-Aguilar, G.A. Antioxidant enrichment and antimicrobial protection of fresh-cut fruits using their own byproducts: Looking for integral exploitation. *J. Food Sci.* **2010**, *75*, R175–R181. [CrossRef] [PubMed]
15. Javed, S.; Mahmood, Z.; Shoaib, A.; Javaid, D.A. Biocidal activity of citrus peel essential oils against some food spoilage bacteria. *J. Med. Plants Res.* **2011**, *5*, 2868–2872.
16. Alsaraf, S.; Hadi, Z.; Al-Lawati, W.M.; Al Lawati, A.A.; Khan, S.A. Chemical composition, in vitro antibacterial and antioxidant potential of Omani Thyme essential oil along with in silico studies of its major constituent. *J. King Saud Univ. Sci.* **2020**, *32*, 1021–1028. [CrossRef]
17. Smigielski, K.; Prusinowska, R.; Stobiecka, A.; Kunicka-Styczyńska, A.; Gruska, R. Biological Properties and Chemical Composition of Essential Oils from Flowers and Aerial Parts of Lavender (*Lavandula angustifolia*). *J. Essent. Oil Bear. Plants* **2018**, *21*, 1303–1314. [CrossRef]
18. Purkait, S.; Bhattacharya, A.; Bag, A.; Chattopadhyay, R.R. Synergistic antibacterial, antifungal and antioxidant efficacy of cinnamon and clove essential oils in combination. *Arch. Microbiol.* **2020**, *202*, 1439–1448. [CrossRef]
19. Meng, F.C.; Zhou, Y.-Q.; Ren, D.; Wang, R.; Wang, C.; Lin, L.G.; Zhang, X.Q.; Ye, W.-C.; Zhang, Q.W. Turmeric: A Review of Its Chemical Composition, Quality Control, Bioactivity, and Pharmaceutical Application. In *Natural and Artificial Flavoring Agents and Food Dyes*; Grumezescu, A.M., Holban, A.M., Eds.; Academic Press: Cambridge, MA, USA, 2018; Chapter 10; pp. 299–350.
20. Bakkali, F.; Averbeck, S.; Averbeck, D.; Idaomar, M. Biological effects of essential oils—A review. *Food Chem. Toxicol.* **2008**, *46*, 446–475. [CrossRef]
21. Ju, J.; Xie, Y.; Guo, Y.; Cheng, Y.; Qian, H.; Yao, W. The inhibitory effect of plant essential oils on foodborne pathogenic bacteria in food. *Crit. Rev. Food Sci. Nutr.* **2019**, *59*, 3281–3292. [CrossRef] [PubMed]

22. Ambrosio, C.M.S.; Ikeda, N.Y.; Miano, A.C.; Saldaña, E.; Moreno, A.M.; Stashenko, E.; Contreras-Castillo, C.J.; Da Gloria, E.M. Unraveling the selective antibacterial activity and chemical composition of citrus essential oils. *Sci. Rep.* **2019**, *9*, 17719. [CrossRef]
23. Deng, W.; Liu, K.; Cao, S.; Sun, J.; Zhong, B.; Chun, J. Chemical Composition, Antimicrobial, Antioxidant, and Antiproliferative Properties of Grapefruit Essential Oil Prepared by Molecular Distillation. *Molecules* **2020**, *25*, 217. [CrossRef]
24. Geraci, A.; Di Stefano, V.; Di Martino, E.; Schillaci, D.; Schicchi, R. Essential oil components of orange peels and antimicrobial activity. *Nat. Prod. Res.* **2017**, *31*, 653–659. [CrossRef]
25. Lockwood, G.B. Techniques for gas chromatography of volatile terpenoids from a range of matrices. *J. Chromatogr. A* **2001**, *936*, 23–31. [CrossRef]
26. Tranchida, P.Q.; Bonaccorsi, I.; Dugo, P.; Mondello, L.; Dugo, G. Analysis of Citrus essential oils: State of the art and future perspectives. A review. *Flavour Fragr. J.* **2012**, *27*, 98–123. [CrossRef]
27. Turek, C.; Stintzing, F.C. Impact of different storage conditions on the quality of selected essential oils. *Food Res. Int.* **2012**, *46*, 341–353. [CrossRef]
28. Turek, C.; Stintzing, F.C. Stability of Essential Oils: A Review. *Compr. Rev. Food Sci. Food Saf.* **2013**, *12*, 40–53. [CrossRef]
29. Dhifi, W.; Bellili, S.; Jazi, S.; Bahloul, N.; Mnif, W. Essential Oils' Chemical Characterization and Investigation of Some Biological Activities: A Critical Review. *Medicines* **2016**, *3*, 25. [CrossRef] [PubMed]
30. Tongnuanchan, P.; Benjakul, S. Essential Oils: Extraction, Bioactivities, and Their Uses for Food Preservation. *J. Food Sci.* **2014**, *79*, R1231–R1249. [CrossRef]
31. Guimaraes, A.; Meireles, L.; Lemos, M.; Guimaraes, M.; Endringer, D.; Fronza, M.; Scherer, R. Antibacterial Activity of Terpenes and Terpenoids Present in Essential Oils. *Molecules* **2019**, *24*, 2471. [CrossRef]
32. Lyu, X.; Lee, J.; Chen, W.N. Potential Natural Food Preservatives and Their Sustainable Production in Yeast: Terpenoids and Polyphenols. *J. Agric. Food Chem.* **2019**, *67*, 4397–4417. [CrossRef]
33. Cutrim, C.S.; Cortez, M.A.S. A review on polyphenols: Classification, beneficial effects and their application in dairy products. *Int. J. Dairy Technol.* **2018**, *71*, 564–578. [CrossRef]
34. Nazzaro, F.; Fratianni, F.; De Martino, L.; Coppola, R.; De Feo, V. Effect of essential oils on pathogenic bacteria. *Pharmaceuticals* **2013**, *6*, 1451–1474. [CrossRef] [PubMed]
35. Gorniak, I.; Bartoszewski, R.; Kroliczewski, J. Comprehensive review of antimicrobial activities of plant flavonoids. *Phytochem. Rev.* **2019**, *18*, 241–272. [CrossRef]
36. Kumar, S.; Pandey, A.K. Chemistry and Biological Activities of Flavonoids: An Overview. *Sci. World J.* **2013**, *2013*, 162750. [CrossRef] [PubMed]
37. Guo, Q.; Liu, K.; Deng, W.; Zhong, B.; Yang, W.; Chun, J. Chemical composition and antimicrobial activity of Gannan navel orange (*Citrus sinensis* Osbeck cv. Newhall) peel essential oils. *Food Sci. Nutr.* **2018**, *6*, 1431–1437. [CrossRef] [PubMed]
38. Uysal, B.; Sozmen, F.; Aktas, O.; Oksal, B.S.; Kose, E.O. Essential oil composition and antibacterial activity of the grapefruit (*Citrus Paradisi*. L) peel essential oils obtained by solvent-free microwave extraction: Comparison with hydrodistillation. *Int. J. Food Sci. Technol.* **2011**, *46*, 1455–1461. [CrossRef]
39. Ou, M.C.; Liu, Y.H.; Sun, Y.W.; Chan, C.F. The Composition, Antioxidant and Antibacterial Activities of Cold-Pressed and Distilled Essential Oils of *Citrus paradisi* and *Citrus grandis* (L.) Osbeck. *Evid.-Based Complement. Altern. Med.* **2015**, *2015*, 804091. [CrossRef] [PubMed]
40. Tao, N.G.; Liu, Y.J. Chemical Composition and Antimicrobial Activity of the Essential Oil from the Peel of Shatian Pummelo (*Citrus Grandis Osbeck*). *Int. J. Food Prop.* **2012**, *15*, 709–716. [CrossRef]
41. Hosni, K.; Zahed, N.; Chrif, R.; Abid, I.; Medfei, W.; Kallel, M.; Brahim, N.B.; Sebei, H. Composition of peel essential oils from four selected Tunisian Citrus species: Evidence for the genotypic influence. *Food Chem.* **2010**, *123*, 1098–1104. [CrossRef]
42. Hou, H.S.; Bonku, E.M.; Zhai, R.; Zeng, R.; Hou, Y.L.; Yang, Z.H.; Quan, C. Extraction of essential oil from Citrus reticulate Blanco peel and its antibacterial activity against *Cutibacterium acnes* (formerly *Propionibacterium acnes*). *Heliyon* **2019**, *5*, e02947. [CrossRef]
43. Peng, Y.; Bishop, K.S.; Quek, S.Y. Compositional analysis and aroma evaluation of Feijoa essential oils from New Zealand grown cultivars. *Molecules* **2019**, *24*, 2053. [CrossRef]
44. Han, Y.; Sun, Z.; Chen, W. Antimicrobial susceptibility and antibacterial mechanism of limonene against *Listeria monocytogenes*. *Molecules* **2020**, *25*, 33. [CrossRef]
45. Fancello, F.; Petretto, G.L.; Zara, S.; Sanna, M.L.; Addis, R.; Maldini, M.; Foddai, M.; Rourke, J.P.; Chessa, M.; Pintore, G. Chemical characterization, antioxidant capacity and antimicrobial activity against food related microorganisms of *Citrus limon* var. pompia leaf essential oil. *LWT Food Sci. Technol.* **2016**, *69*, 579–585. [CrossRef]
46. Swamy, M.K.; Akhtar, M.S.; Sinniah, U.R. Antimicrobial properties of plant essential oils against human pathogens and their mode of action: An updated review. *Evid.-Based Complement. Altern. Med.* **2016**, *2016*, 3012462. [CrossRef] [PubMed]
47. Fisher, K.; Phillips, C. Potential antimicrobial uses of essential oils in food: Is citrus the answer? *Trends Food Sci. Technol.* **2008**, *19*, 156–164. [CrossRef]
48. Wiegand, I.; Hilpert, K.; Hancock, R. Agar and broth dilution methods to determine the minimal inhibitory concentration (MIC) of antimicrobial substance. *Nat. Protoc.* **2008**, *3*, 163–175. [CrossRef] [PubMed]
49. Balouiri, M.; Sadiki, M.; Ibnsouda, S.K. Methods for in vitro evaluating antimicrobial activity: A review. *J. Pharm. Anal.* **2016**, *6*, 71–79. [CrossRef]

50. Al-Fekaiki, D.; Niamah, A.; Al-Sahlany, S. Extraction and identification of essential oil from *Cinnamomum Zeylanicum* barks and study the antibacterial activity. *J. Microbiol. Biotechnol. Food Sci.* **2017**, *7*, 312–316. [CrossRef]
51. Abd-Elwahab, S.M.; El-Tanbouly, N.D.; Moussa, M.Y.; Abdel-Monem, A.R.; Fayek, N.M. Antimicrobial and Antiradical Potential of Four Agro-waste Citrus Peels Cultivars. *J. Essent. Oil-Bear. Plants* **2016**, *19*, 1932–1942. [CrossRef]
52. Bendaha, H.; Bouchal, B.; El Mounsi, I.; Salhi, A.; Berrabeh, M.; El Bellaoui, M.; Mimouni, M. Chemical composition, antioxidant, antibacterial and antifungal activities of peel essential oils of citrus aurantium grown in Eastern Morocco. *Der Pharm. Lett.* **2016**, *8*, 239–245.
53. Gupta, M.; Gularia, P.; Singh, D.; Gupta, S. Analysis of aroma active constituents, antioxidant and antimicrobial activity of *C. Sinensis*, *Citrus limetta* and *C. Limon* fruit peel oil by GC-MS. *Biosci. Biotechnol. Res. Asia* **2014**, *11*, 895–899. [CrossRef]
54. Tao, N.G.; Liu, Y.J.; Zhang, M.L. Chemical composition and antimicrobial activities of essential oil from the peel of bingtang sweet orange (*Citrus sinensis* Osbeck). *Int. J. Food Sci. Technol.* **2009**, *44*, 1281–1285. [CrossRef]
55. Ali, J.; Abbas, S.; Khan, F.A.; Rehman, S.U.; Shah, J.; Rahman, Z.U.; Rahman, I.U.; Paracha, G.M.U.; Khan, M.A.; Shahid, M. Biochemical and antimicrobial properties of *Citrus* peel waste. *Pharmacologyonline* **2016**, *3*, 98–103.
56. Nwachukwu, B.C.; Taiwo, M.O.; Olisemeke, J.K.; Obero, O.J.; Abibu, W.A. Qualitative Properties and Antibacterial Activity of Essential Oil obtained from *Citrus sinensis* Peel on Three Selected Bacteria. *Biomed. J. Sci. Tech. Res.* **2019**, *19*. [CrossRef]
57. Kirbaslar, G.F.; Tavman, A.; Dulger, B.; Turker, G. Antimicrobial activity of Turkish Citrus peel oils. *Pak. J. Bot.* **2009**, *41*, 3207–3212.
58. Van de Vel, E.; Sampers, I.; Raes, K. A review on influencing factors on the minimum inhibitory concentration of essential oils. *Crit. Rev. Food Sci. Nutr.* **2019**, *59*, 357–378. [CrossRef]
59. Faleiro, M.L.; Miguel, M.G.; Ladeiro, F.; Venâncio, F.; Tavares, R.; Brito, J.C.; Figueiredo, A.C.; Barroso, J.G.; Pedro, L.G. Antimicrobial activity of essential oils isolated from Portuguese endemic species of Thymus. *Lett. Appl. Microbiol.* **2003**, *36*, 35–40. [CrossRef] [PubMed]
60. Diep, T.T.; Yoo, M.J.Y.; Pook, C.; Sadooghy-Saraby, S.; Gite, A.; Rush, E. Volatile Components and Preliminary Antibacterial Activity of Tamarillo (*Solanum betaceum* Cav.). *Foods* **2021**, *10*, 2212. [CrossRef] [PubMed]
61. Mandalari, G.; Bennett, R.N.; Bisignano, G.; Trombetta, D.; Saija, A.; Faulds, C.B.; Gasson, M.J.; Narbad, A. Antimicrobial activity of flavonoids extracted from bergamot (*Citrus bergamia* Risso) peel, a byproduct of the essential oil industry. *J. Appl. Microbiol.* **2007**, *103*, 2056–2064. [CrossRef] [PubMed]
62. Hindi, N.; Chabuck, Z. Antimicrobial activity of different aqueous lemon extracts. *J. Appl. Pharm. Sci.* **2013**, *3*, 74–78.
63. El-Hawary, S.; Taha, K.; Abdel-Monem, A.; Kirollos, F.; Mohamed, A. Chemical composition and biological activities of peels and leaves essential oils of four cultivars of *Citrus deliciosa* var. tangarina. *Am. J. Essent. Oils Nat. Prod.* **2013**, *1*, 1–6.
64. Al-Saman, M.A.; Abdella, A.; Mazrou, K.E.; Tayel, A.A.; Irmak, S. Antimicrobial and antioxidant activities of different extracts of the peel of kumquat (*Citrus japonica* Thunb). *J. Food Meas. Charact.* **2019**, *13*, 3221–3229. [CrossRef]
65. Phan, A.D.T.; Chaliha, M.; Sultanbawa, Y.; Netzel, M.E. Nutritional characteristics and antimicrobial activity of Australian grown feijoa (*Acca sellowiana*). *Foods* **2019**, *8*, 376. [CrossRef] [PubMed]
66. Garcia, R.; Alves, E.S.S.; Santos, M.P.; Aquije, G.M.F.V.; Fernandes, A.A.R.; Dos Santos, R.B.; Ventura, J.A.; Fernandes, P.M.B. Antimicrobial activity and potential use of monoterpenes as tropical fruits preservatives. *Braz. J. Microbiol.* **2008**, *39*, 163–168. [CrossRef] [PubMed]
67. Chabuck, Z.A.G.; Al-Charrakh, A.H.; Hindi, N.K.K.; Hindi, S.K.K. Antimicrobial effect of aqueous banana peel extract. *Res. Gate Pharmceutical Sci.* **2013**, *1*, 73–75.
68. Mahmud, S.; Saleem, M.; Siddique, S.; Ahmed, R.; Khanum, R.; Perveen, Z. Volatile components, antioxidant and antimicrobial activity of *Citrus acida* var. sour lime peel oil. *J. Saudi Chem. Soc.* **2009**, *13*, 195–198. [CrossRef]
69. Malviya, S.; Arvind, J.A.; Hettiarachchy, N. Antioxidant and antibacterial potential of pomegranate peel extracts. *J. Food Sci. Technol.* **2014**, *51*, 4132–4137. [CrossRef]
70. Matook, S.M.; Fumio, H. Antibacterial and Antioxidant Activities of Banana (*Musa*, AAA cv. Cavendish) Fruits Peel. *Am. J. Biochem. Biotechnol.* **2005**, *1*, 125–131.
71. Mehmood, T.; Afzal, A.; Anwar, F.; Iqbal, M.; Afzal, M.; Qadir, R. Variations in the Composition, Antibacterial and Haemolytic Activities of Peel Essential Oils from Unripe and Ripened *Citrus limon* (L.) Osbeck Fruit. *J. Essent. Oil-Bear. Plants* **2019**, *22*, 159–168. [CrossRef]
72. Okunowo, W.O.; Oyedeji, O.; Afolabi, L.O.; Matanmi, E. Essential oil of grape fruit (*Citrus paradisi*) peels and its antimicrobial activities. *Am. J. Plant Sci.* **2013**, *4*, 1–9. [CrossRef]
73. Fawole, O.A.; Makunga, N.P.; Opara, U.L. Antibacterial, antioxidant and tyrosinase-inhibition activities of pomegranate fruit peel methanolic extract. *BMC Complement. Altern. Med.* **2012**, *12*, 1–11. [CrossRef]
74. Choi, J.G.; Kang, O.H.; Lee, Y.S.; Chae, H.S.; Oh, Y.C.; Brice, O.O.; Kim, M.S.; Sohn, D.H.; Kim, H.S.; Park, H.; et al. In Vitro and In Vivo Antibacterial Activity of *Punica granatum* peel Ethanol Extract against *Salmonella*. *Evid.-Based Complement. Altern. Med.* **2011**, *2011*, 690518. [CrossRef] [PubMed]
75. Dhanavade, D.M.; Jalkute, D.C.; Ghosh, J.; Sonawane, K. Study Antimicrobial Activity of Lemon (*Citrus lemon* L.) Peel Extract. *Br. J. Pharmacol. Toxicol.* **2011**, *2*, 119–122.
76. Kanatt, S.; Chander, R.; Sharma, A. Antioxidant and antimicrobial activity of pomegranate peel extract improves the shelf life of chicken products. *Int. J. Food Sci. Technol.* **2010**, *45*, 216–222. [CrossRef]

77. Al-Zoreky, N.S. Antimicrobial activity of pomegranate (*Punica granatum* L.) fruit peels. *Int. J. Food Microbiol.* **2009**, *134*, 244–248. [CrossRef]
78. Bajpai, V.K.; Baek, K.-H.; Kang, S.C. Control of *Salmonella* in foods by using essential oils: A review. *Food Res. Int.* **2012**, *45*, 722–734. [CrossRef]
79. Chouhan, S.; Sharma, K.; Guleria, S. Antimicrobial Activity of Some Essential Oils-Present Status and Future Perspectives. *Med. Basel Switz.* **2017**, *4*, 58. [CrossRef]
80. Plesiat, P.; Nikaido, H. Outer membranes of Gram-negative bacteria are permeable to steroid probes. *Mol. Microbiol.* **1992**, *6*, 1323–1333. [CrossRef]
81. Raut, J.S.; Karuppayil, S.M. A status review on the medicinal properties of essential oils. *Ind. Crop. Prod.* **2014**, *62*, 250–264. [CrossRef]
82. Nikaido, H. Prevention of drug access to bacterial targets: Permeability barriers and active efflux. *Science* **1994**, *264*, 382–388. [CrossRef]
83. Zhang, Y.; Liu, X.; Wang, Y.; Jiang, P.; Quek, S. Antibacterial activity and mechanism of cinnamon essential oil against *Escherichia coli* and *Staphylococcus aureus*. *Food Control* **2016**, *59*, 282–289. [CrossRef]
84. Wang, X.; Shen, Y.; Thakur, K.; Han, J.; Zhang, J.G.; Hu, F.; Wei, Z.J. Antibacterial Activity and Mechanism of Ginger Essential Oil against *Escherichia coli* and *Staphylococcus aureus*. *Molecules* **2020**, *25*, 3955. [CrossRef]
85. Bora, H.; Kamle, M.; Mahato, D.K.; Tiwari, P.; Kumar, P. Citrus Essential Oils (CEOs) and Their Applications in Food: An Overview. *Plants* **2020**, *9*, 357. [CrossRef]
86. Cho, T.; Park, S.M.; Yu, H.; Seo, G.; Kim, H.; Kim, S.A.; Rhee, M. Recent Advances in the Application of Antibacterial Complexes Using Essential Oils. *Molecules* **2020**, *25*, 1752. [CrossRef] [PubMed]
87. Kim, J.; Marshall, M.R.; Wei, C.I. Antibacterial Activity of Some Essential Oil Components against Five Foodborne Pathogens. *J. Agric. Food Chem.* **1995**, *43*, 2839–2845. [CrossRef]
88. Griffin, S.G.; Wyllie, S.G.; Markham, J.L.; Leach, D.N. The role of structure and molecular properties of terpenoids in determining their antimicrobial activity. *Flavour Fragr. J.* **1999**, *14*, 322–332. [CrossRef]
89. Zengin, H.; Baysal, A.H. Antibacterial and Antioxidant Activity of Essential Oil Terpenes against Pathogenic and Spoilage-Forming Bacteria and Cell Structure-Activity Relationships Evaluated by SEM Microscopy. *Molecules* **2014**, *19*, 17773–17798. [CrossRef] [PubMed]
90. Togashi, N.; Hamashima, H.; Shiraishi, A.; Inoue, Y.; Takano, A. Antibacterial activities against *Staphylococcus aureus* of terpene alcohols with aliphatic carbon chains. *J. Essent. Oil Res.* **2010**, *22*, 263–269. [CrossRef]
91. Togashi, N.; Inoue, Y.; Hamashima, H.; Takano, A. Effects of Two Terpene Alcohols on the Antibacterial Activity and the Mode of Action of Farnesol against *Staphylococcus aureus*. *Molecules* **2008**, *13*, 3069–3076. [CrossRef]
92. Akiyama, H.; Oono, T.; Huh, W.K.; Yamasaki, O.; Ogawa, S.; Katsuyama, M.; Ichikawa, H.; Iwatsuki, K. Actions of farnesol and xylitol against *Staphylococcus aureus*. *Chemotherapy* **2002**, *48*, 122–128. [CrossRef] [PubMed]
93. Gomes, F.I.A.; Teixeira, P.; Azeredo, J.; Oliveira, R. Effect of farnesol on planktonic and biofilm cells of *Staphylococcus epidermidis*. *Curr. Microbiol.* **2009**, *59*, 118–122. [CrossRef]
94. Jabra-Rizk, M.A.; Meiller, T.F.; James, C.E.; Shirtliff, M.E. Effect of farnesol on *Staphylococcus aureus* biofilm formation and antimicrobial susceptibility. *Antimicrob. Agents Chemother.* **2006**, *50*, 1463–1469. [CrossRef] [PubMed]
95. Saad, N.Y.; Muller, C.D.; Lobstein, A. Major bioactivities and mechanism of action of essential oils and their components. *Flavour Fragr. J.* **2013**, *28*, 269–279. [CrossRef]
96. Lorenzi, V.; Muselli, A.; Bernardini, A.F.; Berti, L.; Pagès, J.M.; Amaral, L.; Bolla, J.M. Geraniol restores antibiotic activities against multidrug-resistant isolates from gram-negative species. *Antimicrob. Agents Chemother.* **2009**, *53*, 2209–2211. [CrossRef]
97. Kumar, M.A.; Devaki, T. Geraniol, a component of plant essential oils-a review of its pharmacological activities. *Int. J. Pharm. Pharm. Sci.* **2015**, *7*, 67–70.
98. Lieutaud, A.; Guinoiseau, E.; Lorenzi, V.; Giuliani, M.C.; Lome, V.; Brunel, J.M.; Luciani, A.; Casanova, J.; Pagès, J.M.; Berti, L.; et al. Inhibitors of antibiotic efflux by AcrAB-TolC in enterobacter aerogenes. *Anti-Infect. Agents* **2013**, *11*, 168–178. [CrossRef]
99. Liu, X.; Cai, J.; Chen, H.; Zhong, Q.; Hou, Y.; Chen, W.; Chen, W. Antibacterial activity and mechanism of linalool against *Pseudomonas aeruginosa*. *Microb. Pathog.* **2020**, *141*, 103980. [CrossRef]
100. Gao, Z.; Van Nostrand, J.D.; Zhou, J.; Zhong, W.; Chen, K.; Guo, J. Anti-*listeria* Activities of Linalool and Its Mechanism Revealed by Comparative Transcriptome Analysis. *Front. Microbiol.* **2019**, *10*, 2947. [CrossRef]
101. Wang, J.-N.; Chen, W.-X.; Chen, R.-H.; Zhang, G.-F. Antibacterial activity and mechanism of limonene against *Pseudomonas aeruginosa*. *J. Food Sci. Technol* **2018**, *39*, 1–5.
102. Herman, A.; Tambor, K.; Herman, A. Linalool Affects the Antimicrobial Efficacy of Essential Oils. *Curr. Microbiol.* **2016**, *72*, 165–172. [CrossRef]
103. Silva, F.; Domingues, F.C. Antimicrobial activity of coriander oil and its effectiveness as food preservative. *Crit. Rev. Food Sci. Nutr.* **2017**, *57*, 35–47. [CrossRef]
104. Silva, F.; Ferreira, S.; Queiroz, J.A.; Domingues, F.C. Coriander (*Coriandrum sativum* L.) essential oil: Its antibacterial activity and mode of action evaluated by flow cytometry. *J. Med Microbiol.* **2011**, *60*, 1479–1486. [CrossRef] [PubMed]
105. Dorman, H.J.D.; Deans, S.G. Antimicrobial agents from plants: Antibacterial activity of plant volatile oils. *J. Appl. Microbiol.* **2000**, *88*, 308–316. [CrossRef] [PubMed]

106. Trombetta, D.; Castelli, F.; Sarpietro, M.G.; Venuti, V.; Cristani, M.; Daniele, C.; Saija, A.; Mazzanti, G.; Bisignano, G. Mechanisms of antibacterial action of three monoterpenes. *Antimicrob. Agents Chemother.* **2005**, *49*, 2474–2478. [CrossRef] [PubMed]
107. Lopez-Romero, J.C.; González-Ríos, H.; Borges, A.; Simões, M. Antibacterial Effects and Mode of Action of Selected Essential Oils Components against *Escherichia coli* and *Staphylococcus aureus*. *Evid.-Based Complement. Altern. Med.* **2015**, *2015*, 795435. [CrossRef] [PubMed]
108. Broniatowski, M.; Mastalerz, P.; Flasiński, M. Studies of the interactions of ursane-type bioactive terpenes with the model of *Escherichia coli* inner membrane—Langmuir monolayer approach. *Biochim. Biophys. Acta* **2015**, *1848*, 469–476. [CrossRef] [PubMed]
109. Dahham, S.S.; Tabana, Y.M.; Iqbal, M.A.; Ahamed, M.B.K.; Ezzat, M.O.; Majid, A.S.A.; Majid, A.M.S.A. The anticancer, antioxidant and antimicrobial properties of the sesquiterpene β-caryophyllene from the essential oil of *Aquilaria crassna*. *Molecules* **2015**, *20*, 11808–11829. [CrossRef] [PubMed]
110. Kim, Y.S.; Park, S.J.; Lee, E.J.; Cerbo, R.M.; Lee, S.M.; Ryu, C.H.; Kim, G.S.; Kim, J.O.; Ha, Y.L. Antibacterial compounds from Rose Bengal-sensitized photooxidation of β-caryophyllene. *J. Food Sci.* **2008**, *73*, C540–C545. [CrossRef]
111. Coman, M.M.; Oancea, A.M.; Verdenelli, M.C.; Cecchini, C.; Bahrim, G.E.; Orpianesi, C.; Cresci, A.; Silvi, S. Polyphenol content and in vitro evaluation of antioxidant, antimicrobial and prebiotic properties of red fruit extracts. *Eur. Food Res. Technol.* **2018**, *244*, 735–745. [CrossRef]
112. Gharib, R.; Najjar, A.; Auezova, L.; Charcosset, C.; Greige-Gerges, H. Interaction of Selected Phenylpropenes with Dipalmitoylphosphatidylcholine Membrane and Their Relevance to Antibacterial Activity. *J. Membr. Biol.* **2017**, *250*, 259–271. [CrossRef]
113. Ferreira, L.; Perestrelo, R.; Caldeira, M.; Câmara, J.S. Characterization of volatile substances in apples from *Rosaceae* family by headspace solid-phase microextraction followed by GC-qMS. *J. Sep. Sci.* **2009**, *32*, 1875–1888. [CrossRef]
114. Matsubara, Y.; Yusa, T.; Sawabe, A.; Iizuka, Y.; Okamoto, K. Structure and Physiological Activity of Phenyl Propanoid Glycosides in Lemon (*Citrus limon* Burm. f.) Peel. *Agric. Biol. Chem.* **1991**, *55*, 647–650.
115. Voo, S.S.; Grimes, H.D.; Lange, B.M. Assessing the Biosynthetic Capabilities of Secretory Glands in *Citrus* Peel. *Plant Physiol.* **2012**, *159*, 81–94. [CrossRef] [PubMed]
116. Atkinson, R.G. Phenylpropenes: Occurrence, Distribution, and Biosynthesis in Fruit. *J. Agric. Food Chem.* **2018**, *66*, 2259–2272. [CrossRef] [PubMed]
117. Qian, W.; Sun, Z.; Wang, T.; Yang, M.; Liu, M.; Zhang, J.; Li, Y. Antimicrobial activity of eugenol against carbapenem-resistant *Klebsiella pneumoniae* and its effect on biofilms. *Microb. Pathog.* **2020**, *139*, 103924. [CrossRef]
118. Cui, H.; Zhang, C.; Li, C.; Lin, L. Antimicrobial mechanism of clove oil on Listeria monocytogenes. *Food Control* **2018**, *94*, 140–146. [CrossRef]
119. Ashrafudoulla, M.; Mizan, M.F.R.; Ha, A.J.-w.; Park, S.H.; Ha, S.-D. Antibacterial and antibiofilm mechanism of eugenol against antibiotic resistance Vibrio parahaemolyticus. *Food Microbiol.* **2020**, *91*, 103500. [CrossRef]
120. Hemaiswarya, S.; Doble, M. Synergistic interaction of eugenol with antibiotics against Gram negative bacteria. *Phytomedicine* **2009**, *16*, 997–1005. [CrossRef]
121. Hyldgaard, M.; Mygind, T.; Piotrowska, R.; Foss, M.; Meyer, R. Isoeugenol has a non-disruptive detergent-like mechanism of action. *Front. Microbiol.* **2015**, *6*, 754. [CrossRef]
122. Marchese, A.; Barbieri, R.; Coppo, E.; Orhan, I.E.; Daglia, M.; Nabavi, S.F.; Izadi, M.; Abdollahi, M.; Nabavi, S.M.; Ajami, M. Antimicrobial activity of eugenol and essential oils containing eugenol: A mechanistic viewpoint. *Crit. Rev. Microbiol.* **2017**, *43*, 668–689. [CrossRef]
123. Auezova, L.; Najjar, A.; Kfoury, M.; Fourmentin, S.; Greige-Gerges, H. Antibacterial activity of free or encapsulated selected phenylpropanoids against Escherichia coli and Staphylococcus epidermidis. *J. Appl. Microbiol.* **2020**, *128*, 710–720. [CrossRef]
124. Albano, M.; Crulhas, B.P.; Alves, F.C.B.; Pereira, A.F.M.; Andrade, B.F.M.T.; Barbosa, L.N.; Furlanetto, A.; Lyra, L.P.d.S.; Rall, V.L.M.; Júnior, A.F. Antibacterial and anti-biofilm activities of cinnamaldehyde against *S. epidermidis*. *Microb. Pathog.* **2019**, *126*, 231–238. [CrossRef] [PubMed]
125. Shen, S.; Zhang, T.; Yuan, Y.; Lin, S.; Xu, J.; Ye, H. Effects of cinnamaldehyde on *Escherichia coli* and *Staphylococcus aureus* membrane. *Food Control* **2015**, *47*, 196–202. [CrossRef]
126. Yin, L.; Chen, J.; Wang, K.; Geng, Y.; Lai, W.; Huang, X.; Chen, D.; Guo, H.; Fang, J.; Chen, Z.; et al. Study the antibacterial mechanism of cinnamaldehyde against drug-resistant *Aeromonas hydrophila* in vitro. *Microb. Pathog.* **2020**, *145*, 104208. [CrossRef] [PubMed]
127. Sharma, S.; Pal, R.; Hameed, S.; Fatima, Z. Antimycobacterial mechanism of vanillin involves disruption of cell-surface integrity, virulence attributes, and iron homeostasis. *Int. J. Mycobacteriol.* **2016**, *5*, 460–468. [CrossRef]
128. Cushnie, T.P.T.; Lamb, A.J. Antimicrobial activity of flavonoids. *Int. J. Antimicrob. Agents* **2005**, *26*, 343–356. [CrossRef]
129. Fathima, A.; Rao, J.R. Selective toxicity of Catechin—A natural flavonoid towards bacteria. *Appl. Microbiol. Biotechnol.* **2016**, *100*, 6395–6402. [CrossRef]
130. Cushnie, T.; Taylor, P.; Nagaoka, Y.; Uesato, S.; Hara, Y.; Lamb, A. Investigation of the antibacterial activity of 3-O-octanoyl-(-)-epicatechin. *J. Appl. Microbiol.* **2008**, *105*, 1461–1469. [CrossRef]
131. Tsuchiya, H.; Iinuma, M. Reduction of membrane fluidity by antibacterial sophoraflavanone G isolated from *Sophora exigua*. *Phytomedicine* **2000**, *7*, 161–165. [CrossRef]
132. Dave, D.; Ghaly, A.E. Meat spoilage mechanisms and preservation techniques: A critical review. *Am. J. Agric. Biol. Sci.* **2011**, *6*, 486–510.

133. Jayasena, D.D.; Jo, C. Essential oils as potential antimicrobial agents in meat and meat products: A review. *Trends Food Sci. Technol.* **2013**, *34*, 96–108. [CrossRef]
134. Lucera, A.; Costa, C.; Conte, A.; Del Nobile, M.A. Food applications of natural antimicrobial compounds. *Front. Microbiol.* **2012**, *3*, 287. [CrossRef] [PubMed]
135. Krasniewska, K.; Kosakowska, O.; Pobiega, K.; Gniewosz, M. The influence of two-component mixtures from Spanish Origanum oil with Spanish Marjoram oil or coriander oil on antilisterial activity and sensory quality of a fresh cut vegetable mixture. *Foods* **2020**, *9*, 1740. [CrossRef] [PubMed]
136. Karagozlu, N.; Ergonul, B.; Ozcan, D. Determination of antimicrobial effect of mint and basil essential oils on survival of *E. coli* O157:H7 and *S. typhimurium* in fresh-cut lettuce and purslane. *Food Control* **2011**, *22*, 1851–1855. [CrossRef]
137. Siddiqua, S.; Anusha, B.A.; Ashwini, L.S.; Negi, P.S. Antibacterial activity of cinnamaldehyde and clove oil: Effect on selected foodborne pathogens in model food systems and watermelon juice. *J. Food Sci. Technol.* **2015**, *52*, 5834–5841. [CrossRef]
138. Huq, T.; Vu, K.D.; Riedl, B.; Bouchard, J.; Lacroix, M. Synergistic effect of gamma (γ)-irradiation and microencapsulated antimicrobials against *Listeria monocytogenes* on ready-to-eat (RTE) meat. *Food Microbiol.* **2015**, *46*, 507–514. [CrossRef]
139. Petrou, S.; Tsiraki, M.; Giatrakou, V.; Savvaidis, I.N. Chitosan dipping or oregano oil treatments, singly or combined on modified atmosphere packaged chicken breast meat. *Int. J. Food Microbiol.* **2012**, *156*, 264–271. [CrossRef]
140. Hsouna, A.B.; Trigui, M.; Mansour, R.B.; Jarraya, R.M.; Damak, M.; Jaoua, S. Chemical composition, cytotoxicity effect and antimicrobial activity of *Ceratonia siliqua* essential oil with preservative effects against *Listeria* inoculated in minced beef meat. *Int. J. Food Microbiol.* **2011**, *148*, 66–72. [CrossRef]
141. Hulankova, R.; Borilova, G.; Steinhauserova, I. Combined antimicrobial effect of oregano essential oil and caprylic acid in minced beef. *Meat Sci.* **2013**, *95*, 190–194. [CrossRef]
142. Vasilatos, G.C.; Savvaidis, I.N. Chitosan or rosemary oil treatments, singly or combined to increase turkey meat shelf-life. *Int. J. Food Microbiol.* **2013**, *166*, 54–58. [CrossRef]
143. Fernandez-Lopez, J.; Viuda-Martos, M. Introduction to the Special Issue: Application of Essential Oils in Food Systems. *Foods* **2018**, *7*, 56. [CrossRef]
144. Kaur, R.; Gupta, T.B.; Bronlund, J.; Kaur, L. The potential of rosemary as a functional ingredient for meat products—A review. *Food Rev. Int.* **2021**, 1–21. [CrossRef]
145. Huang, X.; Lao, Y.; Pan, Y.; Chen, Y.; Zhao, H.; Gong, L.; Xie, N.; Mo, C.-H. Synergistic antimicrobial effectiveness of plant essential oil and its application in seafood preservation: A review. *Molecules* **2021**, *26*, 307. [CrossRef] [PubMed]
146. Carpena, M.; Nuñez-Estevez, B.; Soria-Lopez, A.; Garcia-Oliveira, P.; Prieto, M.A. Essential oils and their application on active packaging systems: A review. *Resources* **2021**, *10*, 7. [CrossRef]
147. Falleh, H.; Ben Jemaa, M.; Saada, M.; Ksouri, R. Essential oils: A promising eco-friendly food preservative. *Food Chem.* **2020**, *330*, 127268. [CrossRef] [PubMed]
148. He, Q.; Guo, M.; Jin, T.Z.; Arabi, S.A.; Liu, D. Ultrasound improves the decontamination effect of thyme essential oil nanoemulsions against *Escherichia coli* O157: H7 on cherry tomatoes. *Int. J. Food Microbiol.* **2021**, *337*, 108936. [CrossRef]
149. Kang, J.-H.; Song, K.B. Inhibitory effect of plant essential oil nanoemulsions against *Listeria monocytogenes*, *Escherichia coli* O157:H7, and *Salmonella typhimurium* on red mustard leaves. *Innov. Food Sci. Emerg. Technol.* **2018**, *45*, 447–454. [CrossRef]
150. Kang, J.-H.; Park, S.-J.; Park, J.-B.; Song, K.B. Surfactant type affects the washing effect of cinnamon leaf essential oil emulsion on kale leaves. *Food Chem.* **2019**, *271*, 122–128. [CrossRef]
151. Yuan, W.; Teo, C.H.M.; Yuk, H.-G. Combined antibacterial activities of essential oil compounds against *Escherichia coli* O157:H7 and their application potential on fresh-cut lettuce. *Food Control* **2019**, *96*, 112–118. [CrossRef]
152. Dai, J.; Li, C.; Cui, H.; Lin, L. Unraveling the anti-bacterial mechanism of Litsea cubeba essential oil against *E. coli* O157:H7 and its application in vegetable juices. *Int. J. Food Microbiol.* **2021**, *338*, 108989. [CrossRef]
153. Krogsgård Nielsen, C.; Kjems, J.; Mygind, T.; Snabe, T.; Schwarz, K.; Serfert, Y.; Meyer, R.L. Antimicrobial effect of emulsion-encapsulated isoeugenol against biofilms of food pathogens and spoilage bacteria. *Int. J. Food Microbiol.* **2017**, *242*, 7–12. [CrossRef]
154. Yoo, J.H.; Baek, K.H.; Heo, Y.S.; Yong, H.I.; Jo, C. Synergistic bactericidal effect of clove oil and encapsulated atmospheric pressure plasma against *Escherichia coli* O157:H7 and *Staphylococcus aureus* and its mechanism of action. *Food Microbiol.* **2021**, *93*, 103611. [CrossRef] [PubMed]
155. Krichen, F.; Hamed, M.; Karoud, W.; Bougatef, H.; Sila, A.; Bougatef, A. Essential oil from pistachio by-product: Potential biological properties and natural preservative effect in ground beef meat storage. *J. Food Meas. Charact.* **2020**, *14*, 3020–3030. [CrossRef]
156. Lin, L.; Mao, X.; Sun, Y.; Rajivgandhi, G.; Cui, H. Antibacterial properties of nanofibers containing chrysanthemum essential oil and their application as beef packaging. *Int. J. Food Microbiol.* **2019**, *292*, 21–30. [CrossRef] [PubMed]
157. Silva, C.d.S.; Figueiredo, H.M.d.; Stamford, T.L.M.; Silva, L.H.M.d. Inhibition of *Listeria monocytogenes* by *Melaleuca alternifolia* (tea tree) essential oil in ground beef. *Int. J. Food Microbiol.* **2019**, *293*, 79–86. [CrossRef] [PubMed]
158. Khaleque, M.A.; Keya, C.A.; Hasan, K.N.; Hoque, M.M.; Inatsu, Y.; Bari, M.L. Use of cloves and cinnamon essential oil to inactivate *Listeria monocytogenes* in ground beef at freezing and refrigeration temperatures. *LWT* **2016**, *74*, 219–223. [CrossRef]
159. Chaichi, M.; Mohammadi, A.; Badii, F.; Hashemi, M. Triple synergistic essential oils prevent pathogenic and spoilage bacteria growth in the refrigerated chicken breast meat. *Biocatal. Agric. Biotechnol.* **2021**, *32*, 101926. [CrossRef]

160. Jayari, A.; El Abed, N.; Jouini, A.; Mohammed Saed Abdul-Wahab, O.; Maaroufi, A.; Ben Hadj Ahmed, S. Antibacterial activity of *Thymus capitatus* and *Thymus algeriensis* essential oils against four food-borne pathogens inoculated in minced beef meat. *J. Food Saf.* **2018**, *38*, e12409. [CrossRef]
161. Kazemeini, H.; Azizian, A.; Adib, H. Inhibition of *Listeria monocytogenes* growth in turkey fillets by alginate edible coating with *Trachyspermum ammi* essential oil nano-emulsion. *Int. J. Food Microbiol.* **2021**, *344*, 109104. [CrossRef]
162. Kazemeini, H.; Azizian, A.; Shahavi, M.H. Effect of chitosan nano-gel/emulsion containing *Bunium Persicum* essential oil and nisin as an edible biodegradable coating on *Escherichia coli* O 157:H 7 in rainbow trout fillet. *J. Water Environ. Nanotechnol.* **2019**, *4*, 343–349.
163. Noori, S.; Zeynali, F.; Almasi, H. Antimicrobial and antioxidant efficiency of nanoemulsion-based edible coating containing ginger (*Zingiber officinale*) essential oil and its effect on safety and quality attributes of chicken breast fillets. *Food Control* **2018**, *84*, 312–320. [CrossRef]
164. Khezri, S.; Khezerlou, A.; Dehghan, P. Antibacterial activity of *Artemisia persica Boiss* essential oil against *Escherichia coli* O157: H7 and *Listeria monocytogenes* in probiotic Doogh. *J. Food Process. Preserv.* **2021**, *45*, e15446. [CrossRef]
165. Valkova, V.; Duranova, H.; Galovicova, L.; Vukovic, N.L.; Vukic, M.; Kacaniova, M. In vitro antimicrobial activity of lavender, mint, and rosemary essential oils and the effect of their vapours on growth of *Penicillium* spp. In a bread model system. *Molecules* **2021**, *26*, 3859. [CrossRef] [PubMed]
166. Ahmed, L.I.; Ibrahim, N.; Abdel-Salam, A.B.; Fahim, K.M. Potential application of ginger, clove and thyme essential oils to improve soft cheese microbial safety and sensory characteristics. *Food Biosci.* **2021**, *42*, 101177. [CrossRef]
167. Santos, M.I.S.; Martins, S.R.; Veríssimo, C.S.C.; Nunes, M.J.C.; Lima, A.I.G.; Ferreira, R.M.S.B.; Pedroso, L.; Sousa, I.; Ferreira, M.A.S.S. Essential oils as antibacterial agents against food-borne pathogens: Are they really as useful as they are claimed to be? *J. Food Sci. Technol.* **2017**, *54*, 4344–4352. [CrossRef]
168. Lages, L.Z.; Radünz, M.; Gonçalves, B.T.; Silva da Rosa, R.; Fouchy, M.V.; de Cássia dos Santos da Conceição, R.; Gularte, M.A.; Barboza Mendonça, C.R.; Gandra, E.A. Microbiological and sensory evaluation of meat sausage using thyme (*Thymus vulgaris*, L.) essential oil and powdered beet juice (*Beta vulgaris* L., Early Wonder cultivar). *LWT* **2021**, *148*, 111794. [CrossRef]
169. Radunz, M.; dos Santos Hackbart, H.C.; Camargo, T.M.; Nunes, C.F.P.; de Barros, F.A.P.; Dal Magro, J.; Filho, P.J.S.; Gandra, E.A.; Radünz, A.L.; da Rosa Zavareze, E. Antimicrobial potential of spray drying encapsulated thyme (*Thymus vulgaris*) essential oil on the conservation of hamburger-like meat products. *Int. J. Food Microbiol.* **2020**, *330*, 108696. [CrossRef]
170. Diarra, M.; Hassan, Y.; Block, G.; Drover, J.; Delaquis, P.; Oomah, B.D. Antibacterial activities of a polyphenolic-rich extract prepared from American cranberry (*Vaccinium macrocarpon*) fruit pomace against *Listeria spp*. *LWT* **2020**, *123*, 109056. [CrossRef]
171. Tamkutė, L.; Gil, B.M.; Carballido, J.R.; Pukalskienė, M.; Venskutonis, P.R. Effect of cranberry pomace extracts isolated by pressurized ethanol and water on the inhibition of food pathogenic/spoilage bacteria and the quality of pork products. *Food Res. Int.* **2019**, *120*, 38–51. [CrossRef]
172. Khanjari, A.; Bahonar, A.; Noori, N.; Siahkalmahaleh, M.R.; Rezaeigolestani, M.; Asgarian, Z.; Khanjari, J. In vitro antibacterial activity of *Pimpinella anisum* essential oil and its influence on microbial, chemical, and sensorial properties of minced beef during refrigerated storage. *J. Food Saf.* **2019**, *39*, e12626. [CrossRef]
173. Sojic, B.; Pavlic, B.; Ikonić, P.; Tomovic, V.; Ikonic, B.; Zekovic, Z.; Kocic-Tanackov, S.; Jokanovic, M.; Skaljac, S.; Ivic, M. Coriander essential oil as natural food additive improves quality and safety of cooked pork sausages with different nitrite levels. *Meat Sci.* **2019**, *157*, 107879. [CrossRef]
174. Zhang, X.; Wang, H.; Li, X.; Sun, Y.; Pan, D.; Wang, Y.; Cao, J. Effect of cinnamon essential oil on the microbiological and physiochemical characters of fresh Italian style sausage during storage. *Anim. Sci. J.* **2019**, *90*, 435–444. [CrossRef]
175. Gniewosz, M.; Stobnicka, A. Bioactive components content, antimicrobial activity, and foodborne pathogen control in minced pork by cranberry pomace extracts. *J. Food Saf.* **2018**, *38*, e12398. [CrossRef]
176. Aminzare, M.; Tajik, H.; Aliakbarlu, J.; Hashemi, M.; Raeisi, M. Effect of cinnamon essential oil and grape seed extract as functional-natural additives in the production of cooked sausage-impact on microbiological, physicochemical, lipid oxidation and sensory aspects, and fate of inoculated *Clostridium perfringens*. *J. Food Saf.* **2018**, *38*, e12459. [CrossRef]
177. Ed-Dra, A.; Rhazi Filali, F.; Bou-Idra, M.; Zekkori, B.; Bouymajane, A.; Moukrad, N.; Benhallam, F.; Bebtayeb, A. Application of *Mentha suaveolens* essential oil as an antimicrobial agent in fresh turkey sausages. *J. Appl. Biol. Biotechnol.* **2018**, *6*, 7–12.
178. Boskovic, M.; Djordjevic, J.; Ivanovic, J.; Janjic, J.; Zdravkovic, N.; Glisic, M.; Glamoclija, N.; Baltic, B.; Djordjevic, V.; Baltic, M. Inhibition of *Salmonella* by thyme essential oil and its effect on microbiological and sensory properties of minced pork meat packaged under vacuum and modified atmosphere. *Int. J. Food Microbiol.* **2017**, *258*, 58–67. [CrossRef] [PubMed]
179. Blanco-Lizarazo, C.M.; Betancourt-Cortés, R.; Lombana, A.; Carrillo-Castro, K.; Sotelo-Díaz, I. *Listeria monocytogenes* behaviour and quality attributes during sausage storage affected by sodium nitrite, sodium lactate and thyme essential oil. *Food Sci. Technol. Int.* **2017**, *23*, 277–288. [CrossRef] [PubMed]
180. Al-Sahlany, S.T.G. Effect of *Mentha piperita* essential oil against *Vibrio* spp. isolated from local cheeses. *Pak. J. Food Sci.* **2016**, *26*, 65–71.
181. European Commission. Regulation (EC) No 1334/2008 of the European Parliament and of the Council of 16 December 2008 on Flavourings and Certain Food Ingredients with Flavouring Properties for Use in and on Foods and Amending Council Regulation (EEC) No 1601/91, Regulations (EC) No 2232/96 and (EC) No 110/2008 and Directive 2000/13/EC. 2008. Available online: https://eur-lex.europa.eu/eli/reg/2008/1334/oj (accessed on 1 January 2022).

182. Ribeiro-Santos, R.; Andrade, M.; Melo, N.R.d.; Sanches-Silva, A. Use of essential oils in active food packaging: Recent advances and future trends. *Trends Food Sci. Technol.* **2017**, *61*, 132–140. [CrossRef]
183. Benkhaira, N.; Koraichi, S.I.; Fikri-Benbrahim, K. In vitro methods to study antioxidant and some biological activities of essential oils: A review. *Biointerface Res. Appl. Chem.* **2022**, *12*, 3332–3347.
184. Yang, T.; Qin, W.; Zhang, Q.; Luo, J.; Lin, D.; Chen, H. Essential-oil capsule preparation and its application in food preservation: A review. *Food Rev. Int.* **2022**, 1–35. [CrossRef]
185. Roda, R.; Taboada-Rodríguez, A.; Valverde-Franco, M.; Marín-Iniesta, F. Antimicrobial Activity of Vanillin and Mixtures with Cinnamon and Clove Essential Oils in Controlling *Listeria monocytogenes* and *Escherichia coli* O157:H7 in Milk. *Food Bioprocess Technol.* **2010**, *5*, 2120–2131. [CrossRef]
186. Gutierrez, J.; Barry-Ryan, C.; Bourke, P. The antimicrobial efficacy of plant essential oil combinations and interactions with food ingredients. *Int. J. Food Microbiol.* **2008**, *124*, 91–97. [CrossRef] [PubMed]
187. Kyung, K.H. Antimicrobial properties of *Allium* species. *Curr. Opin. Biotechnol.* **2012**, *23*, 142–147. [CrossRef]
188. Ruzauskas, M.; Bartkiene, E.; Stankevicius, A.; Bernatoniene, J.; Zadeike, D.; Lele, V.; Starkute, V.; Zavistanaviciute, P.; Grigas, J.; Zokaityte, E.; et al. The influence of essential oils on gut microbial profiles in pigs. *Animals* **2020**, *10*, 1734. [CrossRef]

Article

In Vitro Potential of Clary Sage and Coriander Essential Oils as Crop Protection and Post-Harvest Decay Control Products

Robin Raveau [1], Joël Fontaine [1], Abir Soltani [2], Jouda Mediouni Ben Jemâa [2], Frédéric Laruelle [1] and Anissa Lounès-Hadj Sahraoui [1,*]

[1] Unit of Environmental Chemistry and Interaction on the Living (UCEIV), University of the Littoral Opal Coast (ULCO), UR 4492, SFR Condorcet FR CNRS 3417, 50 Rue Ferdinand Buisson, 62228 Calais, France; robin.raveau@univ-littoral.fr (R.R.); joel.fontaine@univ-littoral.fr (J.F.); frederic.laruelle@univ-littoral.fr (F.L.)

[2] Laboratory of Biotechnology Applied to Agriculture, National Agricultural Research Institute of Tunisia (INRAT), University of Carthage, Rue Hedi Karray, El Menzah, Tunis 1004, Tunisia; Soltani.abyr@gmail.com (A.S.); joudamediouni1969@gmail.com (J.M.B.J.)

* Correspondence: anissa.lounes@univ-littoral.fr

Citation: Raveau, R.; Fontaine, J.; Soltani, A.; Mediouni Ben Jemâa, J.; Laruelle, F.; Lounès-Hadj Sahraoui, A. In Vitro Potential of Clary Sage and Coriander Essential Oils as Crop Protection and Post-Harvest Decay Control Products. *Foods* **2022**, *11*, 312. https://doi.org/10.3390/foods11030312

Academic Editors: Evaristo Ballesteros, Lisa Pilkington and Siew-Young Quek

Received: 14 December 2021
Accepted: 20 January 2022
Published: 24 January 2022

Publisher's Note: MDPI stays neutral with regard to jurisdictional claims in published maps and institutional affiliations.

Copyright: © 2022 by the authors. Licensee MDPI, Basel, Switzerland. This article is an open access article distributed under the terms and conditions of the Creative Commons Attribution (CC BY) license (https:// creativecommons.org/licenses/by/ 4.0/).

Abstract: Owing to their various application fields and biological properties, natural products and essential oils (EO) in particular are nowadays attracting more attention as alternative methods to control plant pathogens and pests, weeds, and for post-harvest applications. Additionally, to overcome EO stability issues and low persistence of effects, EO encapsulation in β-cyclodextrin (β-CD) could represent a promising avenue. Thus, in this work, the EO distilled from two aromatic plants (*Salvia sclarea* L. and *Coriandrum sativum* L.) have been evaluated in vitro for their antifungal, herbicidal and insecticidal activities, against major plant pathogens and pests of agronomical importance. Both plants were grown on unpolluted and trace-element-polluted soils, so as to investigate the effect of the soil pollution on the EO compositions and biological effects. These EO are rich in oxygenated monoterpenes (clary sage and coriander seeds EO), or aliphatic aldehydes (coriander aerial parts EO), and were unaltered by the soil pollution. The tested EO successfully inhibited the growth of two phytopathogenic fungi, *Zymoseptoria tritici* and *Fusarium culmorum*, displaying IC$_{50}$ ranging from 0.46 to 2.08 g L^{-1}, while also exerting anti-germinative, herbicidal, repellent and fumigant effects. However, no improvement of the EO biological effects was observed in the presence of β-CD, under these in vitro experimental conditions. Among the tested EO, the one from aerial parts of coriander displayed the most significant antifungal and herbicidal effects, while the three of them exerted valuable broad-range insecticidal effects. As a whole, these findings suggest that EO produced on polluted areas can be of great interest to the agricultural area, given their faithful chemical compositions and valuable biological effects.

Keywords: essential oils; aromatic plants; antifungal; anti-germinative; herbicidal; insecticidal

1. Introduction

Historically used in traditional medicine, essential oils (EO) are these days raising great interest, owing to their diverse application fields [1–3]. Made of a mixture of volatile compounds, up to 100, and synthesized by all aromatic plant parts as secondary metabolites [4,5], EO were recently outlined for their interest in the preservation of food quality and flavor [6]. They also have received increasing attention as potential alternatives to commercial pesticides in crop protection, given their promising biological properties against plant pathogens, pests and weeds [1,5].

Among fungal phytopathogens, *Fusarium culmorum* and *Zymoseptoria tritici* are of major importance, responsible for *Septoria tritici* blotch and *Fusarium* head blight on cereals, respectively, whose damage on host plant are considerable, and may cause yield losses up to 50% [7,8]. *Fusarium* spp. are also known to produce a wide range of mycotoxins, secondary metabolites that may be highly toxic to human and animal health [8,9]. The control of both *F. culmorum* and *Z. tritici* is mostly achieved through the use of triazole fungicides [8]. However, resistance levels to triazoles have significantly increased since their marketing authorization, and hence compromise their reliability [7,8]. Similarly, insect pests cause significant losses in terms of quantity and quality of the products, in field or during postharvest storage. This is the case for the silverleaf whitefly, *Bemisia tabaci* Genn. (Hemiptera: Aleyrodidae), the lesser grain borer, *Rhyzopertha dominica* F. (Coleoptera: Bostrychidae), or the Mediterranean flour moth, *Ephestia kuehniella* Zeller (Lepidoptera: Pyralidae), who are recognized worldwide as some of the most destructive pests on several economically important crops [10–12]. Their control is also mainly achieved through the use of chemical insecticides, displaying resistance phenomena in pest populations [10,12]. Notably, *B. tabaci* has been identified as resistant to a wide number of systemic insecticides, such as organophosphates, synthetic pyrethroids or neonicotinoids [12,13]. Moreover, their use may lead to detrimental effects on beneficial insects, as well [10,14]. From a wider perspective, the excessive and inappropriate use of pesticides is controversial, because of their noxious impact on both environmental and human health [5,15]. As part of an integrated pest management system, the use of natural products, including EO, considered as biocontrol tools, is then greatly encouraged [16,17], especially in the current context, where regulatory restrictions lead to the withdrawal of several commercial products [15,18]. Owing to their relatively low toxicity for humans and animals, EO are moreover registered as "Generally Recognized As Safe (GRAS)" products by the Food and Drug Administration, and regarded as less harmful to both environmental and human health, in comparison with commercial pesticides [19,20]. They also receive increasing public support, considered as eco-friendly products [4,20].

Nevertheless, EO are highly volatile, which could be of great interest for the reduction of the residues, as well as for postharvest applications [4]. Yet, this appears prohibitive for field applications, due to stability issues, and a short persistency of the biological effects over time [5]. To tackle these problems, the use of appropriate EO formulations could offer a promising tool [4,5,21]. In particular, the use of β-cyclodextrins (β-CD), cyclic oligosaccharides able to encapsulate hydrophobic compounds in aqueous solutions, and cited in the Food and Drug Administration's list of Inactive Pharmaceutical Ingredients, may be of great interest to avoid degradation, while maintaining the products' efficiency [19,22].

Within the plant kingdom, Lamiaceae are one of the biggest flowering plant families, comprising a wide number of valuable aromatic species [23,24]. Among them, clary sage (*Salvia sclarea* L.), a biennial aromatic plant species grown all around the world for its high-value EO, has been long known for its use in the perfumery and cosmetic sectors, but also for its applications in medicine [20,25,26]. Coriander (*Coriandrum sativum* L.), another aromatic plant belonging to the Apiaceae family, is an annual herbaceous plant grown all over the world, for the consumption of its green leaves and its seeds, as a spice, or for EO production [27,28]. Both clary sage and coriander EO have drawn attention given their biological effects, in particular antifungal properties against the phytopathogenic fungi *Phoma* spp., *Alternaria alternata* or yeasts, such as *Candida* spp. [26,28,29], and insecticidal effects against the whitefly *Trialeurodes vaporariorum* [26], or several coleoptera species [30–32].

In addition to the production of EO, aromatic plants may also appear as valuable choices within the framework of phytomanagement approaches [33,34]. There is in fact an urgent need to address the issue bound to the presence in ecosystems of inorganic pollutants, such as trace elements (TE), whose pollution extent may exceed 5 million sites worldwide [35]. They are posing serious threats to environmental and human health, as they are non-degradable, tend to accumulate in living organisms, and exhibit toxic effects when their concentration exceeds a certain bearable threshold [33,36]. With the emergence of phytotechnologies as rising tools to mitigate TE-polluted spaces, the capacity of some aromatic plants to tolerate elevated concentrations of TE could be particularly valuable [33,34,37]. Their cultivation on marginal lands, unsuitable for food production, tends to minimize the risk of food-chain contamination, while avoiding competition with feeding agriculture [33,34]. Moreover, one of the major drawbacks to phytotechnologies lies in most cases in the lack of economic profitability [38]. In that regard, the use of aromatic plants grown on polluted soils for the production of EO, which are biosourced products bearing a high added-value, and displaying a content free of TE, could help towards the obtention of an economic profit [34,39,40].

Nonetheless, EO composition is strongly influenced by environmental factors, namely geographical location, sunlight, climatic conditions and soil properties, including the presence of pollutants [26,41]. Notably, the presence of TE in soil could result in modified EO yield, and altered composition and quality [42–44].

Thus, the aim of this work was first to investigate the potential influence of the soil pollution by TE on the chemical composition of the EO distilled from clary sage inflorescences, and from both aerial parts and seeds of coriander, and then on their biological properties in the presence or in the absence of β-CD. The antifungal, anti-germinative and herbicidal potential, as well as the insecticidal activity of the EO, were evaluated against two major phytopathogenic fungi, namely, *F. culmorum* and *Z. tritici*, two plant species commonly used for chemicals' herbicidal assessments, namely *Lactuca sativa* L. and *Lolium perenne* L., and adults of three insect species, namely, *B. tabaci*, *R. dominica* and *E. kuehniella*, so as to explore their potential use as crop protection products.

2. Materials and Methods

2.1. Essential Oils

The EO tested in this study were acquired by steam distillation of coriander (*Coriandrum sativum* L.), and clary sage (*Salvia sclarea* L.), two aromatic plant species grown in situ on two experimental sites: a TE-polluted one, displaying elevated amounts of TE (7, 394 and 443 ppm of Cd, Pb, and Zn, respectively), and an unpolluted one. Their full description, as well as plant physiological data, are available in [34,45]. The distillation of the harvested aromatic plant biomass was realized in collaboration with a private EO-distiller. The steam distillation (14 m^3 distillation unit—saturated water steam, 0.3 bar) was carried out over a three-hour cycle, until no more EO was recovered, under the previously described experimental conditions [20]. Aerial parts of coriander and seeds were harvested at full blossoming, or at seed maturity, respectively, for their distillation. In the same way, clary sage distillation was performed using harvested inflorescences at full blossoming, during its second year of cultivation, when the highest yields are expected [34,46]. EO were stored at 4 °C, in tightly closed brown glass vials, and under modified nitrogen protective atmosphere, until their use.

2.2. Determination of the EO Chemical Composition

EO samples were first diluted in ethyl acetate (ratio 1:200 (v/v)), and then analyzed by electron ionization gas chromatography–mass spectrometry (Shimadzu QP 2010 Ultra), according to the method previously described [20]. Briefly, volatile EO components were separated on a ZB-5MS (Phenomenex—5%-phenyl-arylene/95% dimethylpolysiloxane—10.0 m × 0.10 mm × 0.10 μm) capillary column. The EO solution was then injected in a split mode (0.2 μL; split ratio 1:10). Helium was used as a carrier gas to operate the system, at a constant linear velocity (60 cm s^{-1}). The column temperature was held for 2 min at 60 °C, then programmed to linearly increase to 280 °C, at a constant rate of 40 °C min^{-1}, and remained at 280 °C for 1 min.

Mass spectra were recorded within a mass range of 35.0 to 350 (m/z), at an interface temperature of 280 °C, and an ionisation energy of 70 eV. The EO components were identified by comparison of their retention indices relative to (C8–C30) n-alkanes (Kovats indices), and their obtained mass spectra, with those listed in the NIST (National Institute of Standards and Technology, Gaithersburg, MD, USA), and Wiley 275 computer libraries, as well as those found in the literature [26,27,41,47,48]. Relative percentages of oil constituents were measured from the GC peak areas.

2.3. Biological Activities of the EO

2.3.1. Antifungal Activity

Phytopathogenic Fungal Strains

Antifungal activities of the different EO were evaluated against two major phytopathogenic fungi—*Fusarium culmorum* and *Zymoseptoria tritici*—by using in vitro assays. *F. culmorum* strain was maintained on PDA (Potato-Dextrose-Agar, Condalab, Spain) medium. The hemibiotrophic fungus *Z. tritici* (strain T02596) was conserved at −80 °C in cryopreservation tubes. Five days prior to the assays, the fungus was cultivated on a PDA medium, in order to produce spores [49].

Determination of In Vitro Antifungal Activity against *F. culmorum*

Essential oils' antifungal activity against *F. culmorum* was evaluated by using an in vitro direct contact assay. It was evaluated according to the method previously described [19,50], with slight modifications. A PDA medium (39 g L^{-1}) was first prepared, complemented with 1% (v/v) DMSO (Thermo Fisher Scientific, Illkirch, France), in which EO were mixed at 50 °C, so as to obtain a final scale of 5 EO concentrations, ranging from 0.005 to 1.0% (v/v) of EO in the medium. Discs of *F. culmorum* (0.9 cm) were then cut out from the periphery of a 7-day-old fungal colony, and placed at the center of a 9 cm Petri dish containing the PDA medium complemented with EO. The assay was carried out in the absence, and the presence of β-CD in the medium, at 10 mM. Mycelium radial growth was measured after a seven-day incubation (20 ± 1 °C). The inhibition rate was calculated following Equation (1):

$$\text{Inhibition rate (\%)} = \frac{X_0 - X_i}{X_0} \times 100 \qquad (1)$$

where X_0 and X_i stand for the average diameter of the fungal colony in control and in treatment, respectively.

Aqueous solutions of DMSO 1% (v/v) or β-CD (10 mM), as well as a marketed fungicide, Aviator XPro (prothioconazole—150 g L^{-1} and bixafen—75 g L^{-1}, Stolz, Wailly-Beaucamp, France), were tested as negative and positive controls, respectively. The positive control was evaluated with concentrations ranging from 5×10^{-5} to 0.5% (v/v). Analyses were led in triplicates for each condition.

Determination of In Vitro Antifungal Activity against *Z. tritici*

Essential oils' antifungal activity against *Z. trici* was evaluated by using an in vitro microplate assay, adapted from Fungicide Resistance Committee methods, and similar to the one developed by [4], with a range of eight concentrations (from 0 to 0.8% of EO in the medium) for each and all EO. Briefly, spores of *Z. trici* were collected and placed in a glucose-peptone suspension. The microplates were then inoculated with 60 µL of the calibrated pathogen suspension (2×10^5 spores mL^{-1}). The microplates were then incubated under agitation at 110 rpm (20 ± 1 °C), in darkness, with an incubation time of 6 days, determined according to the pathogen's optimal growth time. The evaluation of its growth was carried out using a spectrometer (620 nm). For each EO concentration, eight replicate wells were used. Additionally, each assay was carried out in triplicate to compare the products. Controls include four non-inoculated wells per EO concentration. Additionally, EO's effects were assessed in the presence and in the absence of β-CD in the culture medium (10 mM). These natural products were compared to a homologated and marketed product: Aviator XPro (prothioconazole—150 g L^{-1} and bixafen—75 g L^{-1}), within the same range of concentrations.

Determination of In Vitro Antifungal Properties of the EO

The half-maximal inhibitory concentration (IC$_{50}$) of EO (expressed in g L^{-1}), required to obtain a fungal pathogens' growth inhibition of 50%, was calculated for all in vitro assays. A graphical interpolation, complemented with a statistical analysis based on a nonlinear regression, were used to calculate the IC$_{50}$ value [51]. The IC$_{50}$ of each of the tested EO was also classified as either fungicidal or fungistatic, considering its effects. The fungistatic or fungicidal nature of EO was tested by observing growth revival of the inhibited mycelial disc, following its transfer on EO-free PDA medium: no mycelial return to growth defined fungicidal effect, whereas fungistatic effect was characterized by a fungal regrowth capacity on the EO-free medium.

2.3.2. Anti-Germinative and Herbicidal Activities

Inhibitory effects on seedlings' emergence and growth of the different EO, were assessed against two plant species, *Lolium perenne* L. (monocotyledon) and *Lactuca sativa* L. (dicotyledon), commonly used for chemicals' herbicidal assessments [52], and listed in the OECD guidelines (2003). An in vitro method was adapted [53–55]. EO aqueous solutions (DMSO 1% (v/v)) were prepared, in the presence or in the absence of β-CD (10 mM), and mixed in an agar non-complemented medium (50 °C), then poured into square Petri dishes (120 × 120 mm). EO concentrations used ranged from 5×10^{-4} up to 0.5% (v/v). Seeds were then placed on the solidified agar medium, in sealed Petri dishes, and incubated for 8 days on a day/night cycle, with a 16 h photoperiod (20 ± 1 °C), and an obscurity period of 8 h (16 ± 1 °C). Glyphosate (isopropylamine salt—360 g L^{-1}) was used as a positive control within the same concentrations' range as the tested EO, while aqueous solutions of DMSO 1% (v/v) or β-CD (10 mM) were tested as negative ones.

After the incubation period, germination rates were evaluated by counting germinated seeds, while root elongation was assessed through an imaging software (ImageJ), by measuring root length [55,56]. The analyses were led in triplicates. Graphical interpolation, complemented with a statistical analysis [51], was used to calculate the IC$_{50}$ values, regarding both germination and root elongation parameters.

2.3.3. Insecticidal Activities

Insect Individuals

Bemisia tabaci adults were collected from a tomato (*Solanum lycopesicum* L.) greenhouse. Adults of *Rhyzoperta dominica* were kept on whole wheat, while *Epnestia kuehniella* adults were reared on wheat flour. Insects were maintained at 25 ± 1 °C and 65 ± 5% relative humidity. Both female and male adult insects were used for bioassays.

Repellency Bioassay

Repellency bioassays for the different tested EO were carried out according to the experimental methods previously described [57], at 25 ± 1 °C and 65 ± 5% relative humidity. For that, Whatman filter papers (8 cm diameter) were cut in half. Test solutions were prepared by dissolving 0.4, 1 and 2.5 µL of EO in 1 mL acetone. Each solution was applied to half of the filter paper discs, using a micropipette. The other half of the filter paper was treated with acetone only, as a control. The treated and control half discs were then air-dried under a fan, in order to evaporate the totality of the solvent. Treated and untreated halves were attached to their opposites, using adhesive tape, and placed in Petri dishes. Twenty male and female adult insects were then released at the center of each filter paper disc. Parafilm was used to seal the dishes. Three replicates were used for each concentration, and for each EO. The number of insects in the treated and untreated halves was recorded after 1, 3, 5 and 24 h. Three trials were made for each concentration, and tested by applying the χ^2 test for homogeneity ratio (1:1). Numbers of *R. dominica*, *E. kuehniela* and *B. tabaci* adults present on both treated and untreated portions of the experimental paper halves were recorded at different times of exposure. Percentage Repellency (PR) was calculated according to the following formula [58]:

$$\text{Percentage repellency } (\%) = [Nc - Nt \,(Nc + Nt)] \times 100 \quad (2)$$

where Nc and Nt stand for the number of insects on the untreated area and on the treated area, respectively, after various exposure times.

The data were also expressed as RC_{50} values, corresponding to the concentration that repelled 50% of the exposed insects. Three replicates were observed for each EO, at the different exposure times. Replicates were also used for each EO concentration. Comparison was made between the mean number of treated and untreated insects.

Fumigation Bioassay

The toxicity of the three EO by fumigation was tested in Plexiglas bottles of 38 mL, in which 10 *R. dominica*, *E. kuehniella* or *B. tabaci* adults were released. Filter paper was cut into 2 cm in diameter pieces, and impregnated with the different EO concentrations 9.09, 22.72 and 56.81 $\mu L\ L^{-1}$. Caps were tightly screwed on the vials. Mortality was recorded after 2, 4, 6, 24, 36, 48, 72, 96, 120 and 144 h from the start of exposure. Three replicates were done for each EO, and for each concentration. The control did not show any mortality. Results were expressed as median lethal time (LT_{50}), time after which half of a sample population has died, and median lethal concentration (LC_{50}), the chemical concentration that results in the death of 50% of a sample population.

2.4. Statistical Analyses

Statistical analyses were performed using XLSTAT 2018.1.1 (Adinsoft, Paris, France) software and R 3.6.1 [59]. Before any statistical analysis, Shapiro–Wilk and Bartlett tests were performed to verify normality and homoscedasticity assumptions, respectively. When necessary, non-normal data were "square-root" or "log10" transformed.

Regarding antifungal and herbicidal properties, IC_{50} values resulted from non-linear regression analyses from triplicate assays, and were expressed as mean values and standard deviation (mean ± SD). The comparison of IC_{50} values was carried out using two-way analysis of variance (ANOVA), complemented with a *post-hoc* Tukey-HSD (Honestly Significant Difference) test.

For the insecticidal activity, the statistical analysis was performed using SPSS statistical software, version 20.0. When necessary, data were transformed by common logarithm or exponential, to meet the normality assumptions. All obtained values were the mean of three replications, and were expressed as means ± standard error. For the repellent activity, differences between each EO were tested by one-way ANOVA, followed by Duncan test. From the bioassays data, a Probit analysis was further conducted to estimate RC_{50} on one side, and LC_{50} and LT_{50} values on the other.

3. Results

3.1. Determination of the EO Chemical Composition

The GC-MS profiles of the three tested EO, from aerial parts or seeds of coriander, and from clary sage inflorescences, are listed in Table 1. In the EO distilled from aerial parts of coriander, 15 compounds were identified, most of which are aliphatic aldehydes, along with some oxygenated monoterpenes. In contrast, coriander seeds and clary sage EO, in which 11 and 22 compounds were identified, respectively, were particularly rich in terpene compounds (Table 1).

The chromatographic profile showed that linalool was the only compound present in the three EO, from different plants, and plant parts. It was particularly abundant in the EO distilled from coriander seeds, with relative proportions ranging between 76.2 and 80.6%, for the EO distilled from the biomass grown on unpolluted and polluted sites, respectively (Table 1). γ-terpinene also represented a significant proportion of coriander seeds EO (from 7.8 to 8.7%).

Table 1. Chemical composition of the EO from aerial parts or seeds of coriander, and from sage inflorescences, grown on unpolluted or TE-polluted sites. Data are relative percentages of EO compounds, expressed as means ± SD (n = 3). For a same plant part, means followed by an asterisk "*" are significantly different, between polluted and unpolluted conditions, by one-way ANOVA test (α = 0.05).

Experimental Retention Indexes	EO Compounds	Aerial Parts of Coriander		Seeds of Coriander		Sage Inflorescences	
		Unpolluted	Polluted	Unpolluted	Polluted	Unpolluted	Polluted
908	α-pinene	1.2 ± 0.2	1.2 ± 0.2	4.8 ± 0.5 *	3.1 ± 0.3	-	-
944	Camphene	-	-	0.5 ± 0.1	0.4 ± 0.1	-	-
991	β-myrcene	-	-	-	-	1.7 ± 0.3	1.1 ± 0.2
1005	4-carene	-	-	0.1 ± 0.2	0.3 ± 0.5	0.1 ± 0.1	0.1 ± 0.1
1027	Limonene	-	-	1.6 ± 0.1	1.6 ± 0.2	0.3 ± 0.5	0.2 ± 0.2
1034	p-cymene	0.3 ± 0	0.5 ± 0.5	1.6 ± 0	1.3 ± 0.1	-	-
1040	β-phellandrene	-	-	-	-	-	-
1049	Ocimene	-	-	-	-	0.1 ± 0.1	0.1 ± 0
1065	γ-terpinene	1.7 ± 0.2	1.7 ± 0.2	8.7 ± 0.2	7.8 ± 1	0.6 ± 0.1	0.6 ± 0.2
1100	Linalool	26.8 ± 4.3	34.5 ± 4.1	76.2 ± 1	80.6 ± 2.3 *	10.3 ± 0.2	15.4 ± 1 *
1133	Camphor	1.2 ± 0.2	1.5 ± 0.2	3.6 ± 0.1	3.7 ± 0.1	-	-
1193	α-terpineol	-	-	-	-	1.7 ± 0.3	2 ± 0.3
1205	Decanal	7.5 ± 0.8	7.5 ± 0.5	-	-	-	-
1250	Linalyl acetate	-	-	-	-	52.2 ± 1.4	62.7 ± 0.2 *
1274	(Z)-2-decenal	49.1 ± 2	44 ± 5.3	-	-	-	-
1308	Undecanal	-	1.8 ± 1.5 *	-	-	-	-
1371	2-undecenal	1.4 ± 0.3	0.8 ± 0.7	-	-	-	-
1375	α-copaene	-	-	-	-	3.9 ± 0.5 *	2 ± 0.1
1383	Geranyl acetate (cis)	-	-	-	-	1 ± 0.2	0.9 ± 0.2
1386	Geranyl acetate (trans)	-	-	2.1 ± 0.8	1.3 ± 0.9	2.2 ± 0.2	2.2 ± 0.6
1389	β-cubebene	-	-	-	-	0.1 ± 0.1	0.1 ± 0
1414	β-caryophyllene	-	-	-	-	3.4 ± 0.4	2.4 ± 0.3
1420	Dodecanal	0.6 ± 0.1	0.8 ± 0	-	-	-	-
1427	β-copaene	-	-	-	-	1.2 ± 0.2	0.6 ± 0.1
1447	β-farnesene	1.4 ± 0.1 *	0.2 ± 0.2	-	-	0.03 ± 0.1	0.03 ± 0.1
1467	2-dodecenal	5 ± 0.7 *	3.5 ± 0.4	-	-	-	-
1479	Germacrene D	-	-	0.3 ± 0.5 *	-	15.6 ± 1.3 *	7.1 ± 0.3
1484	α-Humulene	-	-	-	-	0.4 ± 0	0.2 ± 0.1
1515	Tridecanal	0.3 ± 0.5 *	-	-	-	-	-
1523	β-cadinene	-	-	-	-	1.3 ± 0.2	0.5 ± 0.1
1551	Germacrene B	-	-	-	-	1.8 ± 0.2	0.9 ± 0
1570	2-tridecenal	3.7 ± 0.2 *	2.3 ± 0.2	-	-	-	-
1580	Caryophyllene oxide	-	-	0.1 ± 0.1 *	-	0.3 ± 0.1	0.2 ± 0
1900	Sclareol oxide	-	-	-	-	0.5 ± 0.1	0.4 ± 0.1
2220	Sclareol	-	-	-	-	0.3 ± 0	0.3 ± 0.1

"-": undetected compound.

In the EO distilled from coriander aerial parts, the other compounds identified with a proportion higher than 5% were decanal (7.5%) and (Z)-2-decenal (44 to 49.1%), whereas linalyl acetate and germacrene D were the other major compounds in clary sage EO (varying between 52.2 and 62.7%, and between 7.1 and 15.6%, respectively).

It should also be noted that, even though the overall composition was highly similar for a same plant species and part, the balance between several compounds varied slightly, between EO distilled from aromatic plants grown on unpolluted and TE-polluted sites. It is notably the case for linalool, linalyl acetate, and germacrene-D in EO from sage inflorescences, for undecanal, 2-dodecenal, or 2-tridecenal in the EO distilled from aerial parts of coriander, and for α-pinene, linalool or β-farnesene in the EO from coriander seeds (Table 1).

3.2. EO Antifungal Activity

3.2.1. In Vitro Antifungal Activity against F. culmorum

Our results have shown that the EO from both coriander and sage presented antifungal properties against the phytopathogenic fungus, *F. culmorum*. The results obtained with coriander aerial parts' EO range of concentrations are provided in Figure S1. It was characterized as fungistatic, since fungal regrowth was observed when the discs containing the fungus were transferred on an EO-free medium. IC_{50} obtained for the direct contact assay ranged from 0.46 to 2.08 g L^{-1}, with no significant difference observed between the EO from a same plant, either aerial parts of coriander, seeds of coriander or clary sage, but cultivated under the different experimental conditions (polluted or unpolluted site—Figure 1). Additionally, EO distilled from either aerial parts or seeds of coriander demonstrated a higher efficiency (displaying lower IC_{50}) than those from sage, with IC_{50} ranging from 0.46 to 0.53 g L^{-1}, and from 1.47 to 2.08 g L^{-1}, for coriander and sage, respectively. On another note, no significant improvement was observed in the presence of β-CD, displaying either similar or higher IC_{50}, in comparison with the β-CD-free condition. In comparison with the positive control, all the obtained IC_{50} were significantly higher, up to 104 times (Figure 1).

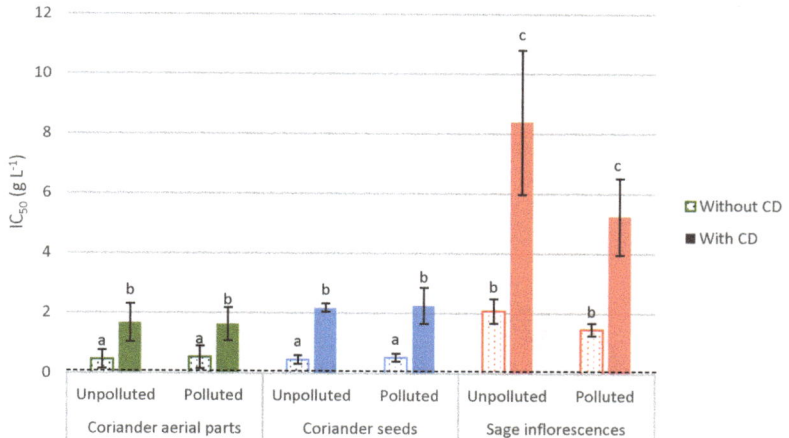

Figure 1. EO's IC_{50} values (g L^{-1}) arising from the antifungal in vitro direct contact bioassay against *F. culmorum*. Values are means ± SD (n = 3). Means followed by the same lowercase letter are not significantly different, by two-way ANOVA comparison (α = 0.05). The positive control (Aviator XPro) value is represented by the black dotted line. All conditions are different from the positive control. IC_{50}: half-maximum inhibitory concentration; CD: cyclodextrins.

3.2.2. In Vitro Antifungal Activity against *Z. tritici*

Our results from the microplate assay against *Z. tritici* have shown that the EO from coriander aerial parts or seeds, and sage, presented fungistatic properties against this phytopathogenic fungus. The results obtained for the in vitro microplate bioassay with the three tested EO are provided in Figure S2. IC_{50} obtained for this assay ranged from 0.001 to 0.08 g L^{-1}, with no significant difference observed, whatever the EO and the plant part it is distilled from (aerial parts or seeds of coriander, sage inflorescences) and the experimental cultivation conditions (polluted or unpolluted sites—Figure 2). No significant improvement was observed in the presence of β-CD, displaying either similar or higher IC_{50} in comparison with the β-CD-free condition. In comparison with the positive control, the IC_{50} obtained for both coriander EO (aerials parts or seeds) and sage were similar in our experimental conditions.

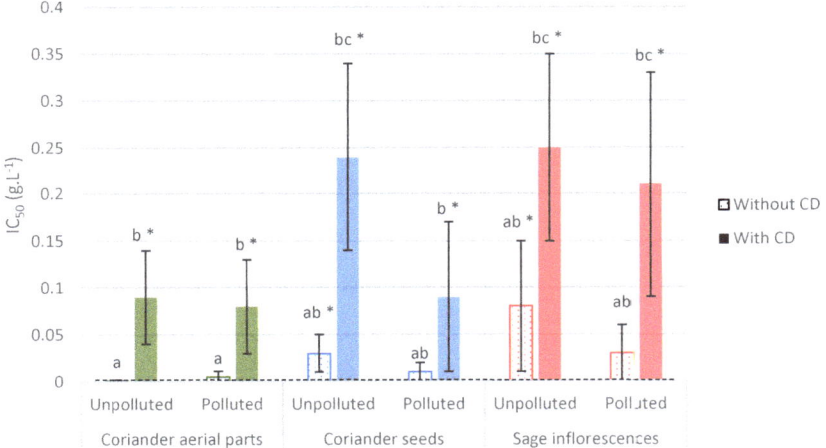

Figure 2. EO's IC_{50} values (g L^{-1}) arising from the antifungal in vitro microplate bioassay against *Z. tritici*. Values are means ± SD (n = 3). Means followed by the same lowercase letter are not significantly different, by two-way ANOVA comparison (α = 0.05). The positive control (Aviator XPro) value is represented by the black dotted line. All conditions different from the positive control are displayed with an asterisk "*". IC_{50}: half-maximum inhibitory concentration; CD: cyclodextrins.

3.3. EO Anti-Germinative and Herbicidal Activities

3.3.1. Seedlings' Emergence Inhibition Bioassay

Our results have shown that all tested EO exerted a significant anti-germinative effect on both lettuce and rye-grass. The obtained IC_{50} ranged from 0.05 to 6.22 g L^{-1}, and from 0.15 to 9.9 g L^{-1}, for EO tested on lettuce, and rye-grass, respectively (Table 2). EO from aerial parts and seeds of coriander have demonstrated a higher efficiency, on both lettuce and rye-grass, than those from sage. Additionally, no difference was found between the EO distilled from a same coriander part, but under the different experimental conditions (polluted or unpolluted soil).

In the presence of β-CD, no significant improvement was demonstrated in our experimental conditions, with even significantly higher effects in the absence of β-CD for the EO of coriander's aerial parts, and of sage, on both tested plants (Table 2). Due to the retention of EO by β-CD, it sometimes resulted in a very limited efficiency at the tested concentrations. Thus, IC_{50} have not been calculated (NC) for several conditions, as it would have resulted in particularly high and inaccurate values. On another note, in comparison with the positive control, only the EO from aerial parts of coriander are in the same range of efficiency on lettuce, whereas all the tested EO are at least as efficient as the control on rye-grass.

3.3.2. Seedlings' Growth Inhibition Bioassay

IC_{50} obtained regarding radicle growth inhibition varied from 0.017 to 1.17 g L^{-1}, and from 0.050 to 0.66 g L^{-1}, for lettuce and rye-grass bioassays, respectively (Table 3). On both lettuce and rye-grass, EO from coriander aerial parts displayed the highest efficiency (Figure S3), in comparison with those from coriander seeds, and from sage. In addition, EO from coriander seeds displayed higher efficiency than those from sage on lettuce. On another note, no difference was visible between the EO distilled from a same plant, but under the different experimental conditions (polluted or unpolluted soil), against both lettuce and rye-grass (Table 3). Additionally, in the presence of β-CD, no significant improvement was obtained in our experimental conditions, with even significantly lower effects in the presence of β-CD (negative effect), in most cases. In comparison with the positive control, the tested EO displayed IC_{50} more than 100 times higher in the case of both lettuce and rye-grass assays.

3.4. EO Insecticidal Activities

The mortality rates of *E. kuehniella*, *B. tabaci* and *R. dominica*, exposed for 24 h to different concentrations of clary sage and coriander EO, are presented in Figure 3.

Data related to the effect of the three tested EO against *E. kuehniella* showed that the EO obtained from clary sage and coriander seeds were toxic at the lowest tested concentration (9.09 µL L^{-1} air), displaying mortality percentages ranging from 3.33 to 16.7% (Figure 3A). The mortality percentage significantly increased with higher EO concentrations, to attain 50% at the highest evaluated concentration (56.81 µL L^{-1} air), for both EO. Conversely, the EO from coriander aerial parts did not exert toxic effect at the lowest concentration, while displaying mortality percentages up to 13.3 and 10%, for the unpolluted and polluted conditions, respectively (Figure 3A). According to the mortality percentages, EO toxicity increased in the following order: coriander seeds EO ≤ clary sage EO ≤ coriander aerial parts EO.

These results also demonstrated that the three tested essential oils had toxic effects against *B. tabaci*, even at the lowest tested concentration (9.09 µL L^{-1} air—Figure 3B). The mortality percentages ranged between 26.7 and 50%, at the lowest and highest tested concentrations, respectively. Similar values were obtained between the three tested EO ($p > 0.05$), whatever the condition (polluted or unpolluted site).

Regarding *R. dominica* mortality, the results indicated that no mortality was caused by clary sage, and coriander aerial parts EO, at the lowest concentration (9.09 µL L^{-1} air). Both EO displayed a similar toxicity pattern, with mortality percentages significantly increasing with higher EO concentrations, to attain 20.0 and 6.7% at the highest concentration (56.81 µL L^{-1} air), for the unpolluted and polluted conditions, respectively (Figure 3C). Moreover, lower mortality percentages were obtained with the EO obtained from the plants grown in polluted conditions (F = 72.64; $p \leq 0.001$). Additionally, in comparison with these 2 EO, significantly higher toxic effects were obtained with the EO from coriander seeds, displaying lethal effects at the lowest tested concentration (ranging from 6.7 to 16.7%), increasing to up to 100% at the highest concentration (56.81 µL L^{-1} air—Figure 3C).

The statistical analysis showed highly significant differences between EO, especially between clary sage or coriander aerial parts EO on one side, and coriander seeds EO on the other, for *R. dominica* (F = 4194.41; $p \leq 0.001$), and *B. tabaci* (F = 435; $p \leq 0.001$), or between coriander aerial parts on one side, and clary sage and coriander seeds EO on the other, for *E. kuehniella* (F = 583.65; $p \leq 0.001$). Moreover, whatever the EO, the concentration had a significant effect on the mortality percentages of the three tested insects, and especially *E. kuehniella* and *R. dominica* (F = 3693.01; $p \leq 0.001$).

Table 2. EO's IC$_{50}$ (g L^{-1}) resulting from the seedlings' emergence inhibition bioassay against lettuce and rye-grass. Values are means ± SD (n = 3).

Bioassay		Aerial Parts of Coriander		Seeds of Coriander		Sage Inflorescences		Positive Control
		Without β-CD	With β-CD	Without β-CD	With β-CD	Without β-CD	With β-CD	
Lettuce	Unpolluted	0.05 ± 0.01 a	603 ± 138 b	0.56 ± 0.12 c	1.87 ± 0.17 d	6.22 ± 2.72 e	NC	0.014 ± 0.006 a
	Polluted	0.08 ± 0.01 a	604 ± 107 b	0.73 ± 0.24 c	1.86 ± 0.02 d	4.21 ± 1.08 e	NC	
Rye-grass	Unpolluted	0.15 ± 0.03 a'	500 ± 317 b'	0.60 ± 0.06 c'	1.73 ± 0.26 d'	2.6 ± 0.39 d'	3093 ± 975 f'	36.5 ± 15.1 e'
	Polluted	0.16 ± 0.05 a'	782 ± 679 b'	0.74 ± 0.13 c'	1.80 ± 0.26 d'	9.9 ± 8.3 e'	1273 ± 555 f'	

IC$_{50}$: inhibitory concentration; NC: not calculable; Positive control: glyphosate. The different letters are the result from a two-way ANOVA comparison (α = 0.05), between results obtained in free β-CD condition, and in the presence of β-CD. Means followed by the same letter—without and with apostrophe for lettuce and rye-grass assays, respectively—do not significantly differ.

Table 3. EO's IC$_{50}$ (g L^{-1}) resulting from the growth inhibition bioassay against lettuce and rye-grass. Values are means ± SD (n = 3).

Bioassay		Aerial Parts of Coriander		Seeds of Coriander		Sage Inflorescences		Positive Control
		Without β-CD	With β-CD	Without β-CD	With β-CD	Without β-CD	With β-CD	
Lettuce	Unpolluted	0.017 ± 0.001 a	0.31 ± 0.07 b	0.28 ± 0.05 b	2.09 ± 0.14 c	1.16 ± 0.45 c	3.96 ± 2.89 c	0.0001 ± 0.0001 d
	Polluted	0.028 ± 0.010 a	0.19 ± 0.08 b	0.29 ± 0.07 b	1.90 ± 0.36 c	1.17 ± 0.56 c	2.49 ± 0.50 c	
Rye-grass	Unpolluted	0.050 ± 0.014 a'	0.84 ± 0.40 b'	0.25 ± 0.07 c'	1.93 ± 0.33 d'	0.66 ± 0.14 bc'	3.09 ± 2.2 d'	0.0016 ± 0.0006 e'
	Polluted	0.053 ± 0.010 a'	0.94 ± 0.80 b'	0.25 ± 0.04 c'	1.41 ± 0.24 d'	0.50 ± 0.04 bc'	2.52 ± 1.52 d'	

IC$_{50}$: inhibitory concentration; NC: not calculable; Positive control: glyphosate. The different letters are the result from a two-way ANOVA comparison (α = 0.05), between results obtained in free β-CD condition, and in the presence of β-CD. Means followed by the same letter—without and with apostrophe for lettuce and rye-grass assays, respectively—do not significantly differ.

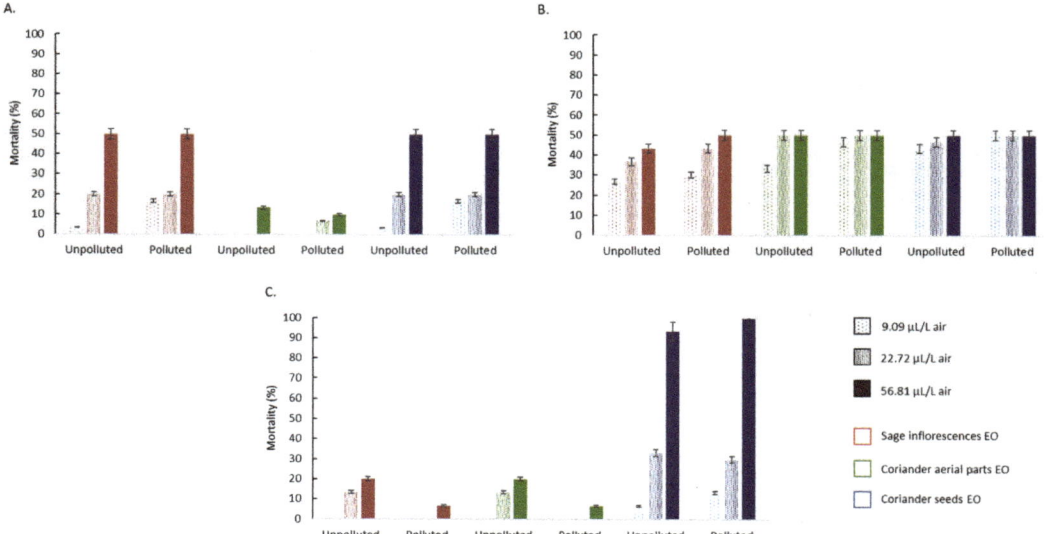

Figure 3. Mortality rates (%) of *E. kuehniella* (**A**), *B. tabaci* (**B**) and *R. dominica* (**C**) adults, exposed during 24 h to three EO (clary sage, coriander aerial parts and coriander seeds) at different concentrations (n = 3).

3.4.1. Repellent Activity

The results in terms of EO repellent activity against *E. kuehniella*, *B. tabaci* and *R. dominica*, have been evaluated using impregnated filter paper test, and are illustrated in Table 4. Whatever the condition (unpolluted or polluted sites), and the target insect, the three tested EO exhibited an important repellent activity (F = 9.01; $p \leq 0.001$), ranging from 13.3 to 66.7%, and dependent upon EO, and concentration ($p < 0.05$). In this test, the strongest repellent activity against *E. kuehniella* was caused by the EO distilled from the aerial parts of coriander, at the highest tested concentration (0.1 µL cm^{-2}—66.7%), while clary sage EO showed the highest repellent activity against *R. dominica*, up to 63.3% (Table 4). Against *B. tabaci*, the three tested EO revealed similar efficiencies (F = 1.01; $p = 0.36$).

Furthermore, these results showed that the origin of the EO (polluted or unpolluted sites) had no effect on EO repellent potential (F = 1.42; $p = 0.24$), and that the effects of increasing EO concentrations are particularly marked against *E. kuehniella* (F = 65.31; $p \leq 0.001$) and *B. tabaci* (F = 12.61; $p \leq 0.001$).

Regarding the median repellent concentrations (RC$_{50}$), the three tested EO showed good repellent activity against the three target insects, displaying RC$_{50}$ values ranging from 2.61 to 3.80 µL cm^{-2}, from 2.61 to 3.77 µL cm^{-2}, and from 0.07 to 0.16 µL cm^{-2}, against *E. kuehniella*, *B. tabaci* and *R. dominica*, respectively (Table S1). It should be noted that *R. dominica* displayed the highest sensitivity to the EO treatments, whatever the EO, and the condition (polluted or unpolluted sites). Furthermore, whatever the insect, the tested EO displayed similar ranges of efficiency, after 24 h of exposure, regardless of the condition (Table S1).

Table 4. Percentage repellency of the three different EO (sage and aerial parts and seeds of coriander), from the two experimental plots—after 24 h of exposure—against E. kuehniella, B. tabaci and R. dominica adults. Values are means ± SE (n = 3).

Insect Species		Aerial Parts of Coriander		Seeds of Coriander		Sage Inflorescences	
		Unpolluted	Polluted	Unpolluted	Polluted	Unpolluted	Polluted
E. kuehniella	0.016 µL cm^{-2}	13.3 ± 0.6	13.3 ± 1.1	10.0 ± 1.3	12.5 ± 0.9	6.7 ± 1.1	10.0 ± 0.6
	0.04 µL cm^{-2}	26.7 ± 5.8	36.7 ± 3.1	28.9 ± 10.2	18.9 ± 1.7	26.7 ± 2.0	33.4 ± 2.8
	0.1 µL cm^{-2}	66.7 ± 7.9	46.7 ± 11.5	50.0 ± 9.6	42.2 ± 11.8	43.4 ± 1.8	46.7 ± 0.3
B. tabaci	0.016 µL cm^{-2}	13.3 ± 1.6	13.3 ± 0.7	6.7 ± 2.1	13.3 ± 0.5	6.8 ± 0.5	6.8 ± 1.7
	0.04 µL cm^{-2}	20.0 ± 4.6	26.7 ± 3.1	20.0 ± 0	26.7 ± 5.7	26.7 ± 0.8	26.7 ± 2.1
	0.1 µL cm^{-2}	40.0 ± 6.7	53.3 ± 5.8	40.0 ± 5.0	40.0 ± 5.7	40.0 ± 2.5	53.3 ± 3.1
R. dominica	0.016 µL cm^{-2}	30.0 ± 1.7	16.7 ± 1.3	16.7 ± 3.4	13.3 ± 4.9	16.7 ± 1.7	16.7 ± 4.0
	0.04 µL cm^{-2}	23.3 ± 4.8	23.3 ± 4.9	23.3 ± 0.7	30.0 ± 4.1	16.7 ± 2.9	53.3 ± 3.3
	0.1 µL cm^{-2}	33.4 ± 3.3	33.3 ± 4.7	36.7 ± 2.2	43.3 ± 2.3	63.3 ± 2.6	56.7 ± 0.9

3.4.2. Fumigant Toxicity

The results in terms of fumigant activity were expressed as both median lethal time (LT_{50}), and median lethal concentration (LC_{50}) values.

Regarding *E. kuehniella*, the three tested EO exerted a significant activity by fumigation. LC_{50} values ranged between 3.0 and 5.2 µL L^{-1}, with mean median lethal times estimated between 87 and 141 h (Table 5). Moreover, no significant difference was observed between LC_{50} values, whatever the EO ($p > 0.05$). However, LC_{50} values were significantly different between the EO distilled from aerial parts of coriander, originating from the unpolluted (3.5 µL L^{-1}), and polluted (5.2 µL L^{-1}) sites.

Table 5. LT_{50} (h) and LC_{50} (µL L^{-1}) values resulting from the fumigation bioassay against E. kuehniella adults, for the three different tested EO, originating from the two experimental plots (n = 3).

Distilled Plant Part		Concentration (µL L^{-1})	LT_{50} (h)	LC_{50}	χ^2	Slope ± SE	p
Sage inflorescences	Unpolluted	9.09	101.6	3.0	0.24	0.9 ± 0.1	0.02
		22.72	99.8				
		56.81	121.0				
	Polluted	9.09	123.5	3.2	4.40	0.5 ± 0.1	0.04
		22.72	97.4				
		56.81	41.2				
Aerial parts of coriander	Unpolluted	9.09	121.8	3.5	0.51	2.2 ± 1.4	0.01
		22.72	113.3				
		56.81	187.7				
	Polluted	9.09	136.9	5.2	3.41	0.5 ± 0.03	0.01
		22.72	136.2				
		56.81	131.3				
Seeds of coriander	Unpolluted	9.09	135.5	3.5	11.80	0.7 ± 0.1	0.01
		22.72	103.2				
		56.81	102.4				
	Polluted	9.09	104.2	3.8	1.49	0.5 ± 0.02	0.01
		22.72	126.8				
		56.81	102.8				

LT_{50}: median lethal time; LC_{50}: median lethal concentration; SE: standard error.

Against *B. tabaci*, the data related to LT_{50} values ranged from 25.5 to 37.6 h. All the three tested EO have demonstrated a fumigant lethal potential, at low concentrations ranging from 2.7 to 3.7 µL L^{-1} of EO (Table 6). No significant difference was observed between either LT_{50} or LC_{50} values, whatever the EO, and the plant growing conditions (unpolluted or polluted sites).

Table 6. LT_{50} (h) and LC_{50} (µL L^{-1}) values resulting from the fumigation bioassay against *B. tabaci* adults, for the three different tested EO, originating from the two experimental plots (n = 3).

Distilled Plant Part		Concentration (µL L^{-1})	LT_{50} (h)	LC_{50}	χ^2	Slope ± SE	p
Sage inflorescences	Unpolluted	9.09 22.72 56.81	37.6 33.9 35.4	3.7	0.16	0.1 ± 0.1	0.01
	Polluted	9.09 22.72 56.81	36.2 29.4 32.8	2.9	0.30	0.3 ± 0.1	0.01
Aerial parts of coriander	Unpolluted	9.09 22.72 56.81	36.2 27.1 25.5	2.7	2.03	0.2 ± 0.1	0.01
	Polluted	9.09 22.72 56.81	31.6 28.4 26.5	2.7	0.06	0.04 ± 0.01	0.04
Seeds of coriander	Unpolluted	9.09 22.72 56.81	36.0 27.0 25.5	3.6	0.01	0.3 ± 0.1	0.05
	Polluted	9.09 22.72 56.81	29.4 25.5 25.5	2.7	0.11	0.1 ± 0.01	0.05

LT_{50}: median lethal time; LC_{50}: median lethal concentration; SE: standard error.

Concerning *R. dominica*, LC_{50} values ranged between 2.2 and 4.1 µL L^{-1}, while LT_{50} values were measured between 19.8 and 123.1 h (Table 7). No significant difference was observed regarding LC_{50} values, between EO from clary sage inflorescences and aerial parts of coriander ($p > 0.05$), and whatever the plant growing conditions (unpolluted or polluted sites—$p > 0.05$). Nonetheless, the EO distilled from seeds of coriander was more toxic against *R. dominica*, displaying lower LC_{50}, ranging between 2.2 and 2.9 µL L^{-1}.

Table 7. LT_{50} (h) and LC_{50} (µL L^{-1}) values resulting from the fumigation bioassay against *R. dominica* adults, for the three different tested EO, originating from the two experimental plots (n = 3).

Distilled Plant Part		Concentration (µL L^{-1})	LT_{50} (h)	LC_{50}	χ^2	Slope ± SE	p
Sage inflorescences	Unpolluted	9.09 22.72 56.81	84.8 60.4 123.1	4.1	5.73	0.7 ± 0.2	0.01
	Polluted	9.09 22.72 56.81	108.5 68.6 11.2	3.8	0.05	2.0 ± 1.6	0.04
Aerial parts of coriander	Unpolluted	9.09 22.72 56.81	84.8 60.4 123.1	4.1	6.04	0.7 ± 0.2	0.02
	Polluted	9.09 22.72 56.81	108.5 68.6 118.7	3.7	0.05	2.0 ± 1.7	0.03
Seeds of coriander	Unpolluted	9.09 22.72 56.81	19.8 100.0 97.5	2.2	5.69	1.5 ± 0.1	0.01
	Polluted	9.09 22.72 56.81	103.1 84.2 62.9	2.9	6.87	0.2 ± 0.1	0.01

LT_{50}: median lethal time; LC_{50}: median lethal concentration; SE: standard error.

As a whole, it should be noted that the three target insects displayed similar ranges of susceptibility to the different EO, tested by fumigation.

4. Discussion

Natural compounds from plants, including EO, may be efficient alternatives to the conventional pesticides, especially in integrated approaches. First of all, the possible influence of the soil pollution by TE, on the EO chemical compositions, was assessed.

Then, the potential of EO obtained from clary sage (inflorescences) and coriander (aerial parts and seeds), regarding their antifungal, anti-germinative, herbicidal, and insecticidal activities, was investigated in vitro.

4.1. Effect of Soil Pollution on the Chemical Compositions of EO Distilled from Coriander and Clary Sage

In this study, the chemical composition of the EO distilled from coriander aerial parts was characterized by significant proportions of 2-decenal (between 44 and 49%), linalool (up to 35%), decanal, 2-dodecenal, and 2-tridecenal, among 15 different detected aromatic compounds. It is hence mostly composed of aliphatic aldehydes, which is consistent with previous investigations [28,60,61]. Some of these previous reports have shown that the EO chemical composition, and that of coriander EO in particular, was dependent upon the plant part that was used for the distillation [28,60,62]. Thus, it is not surprising that the EO distilled from seeds of coriander displayed a significantly different composition from the one distilled from its aerial parts, and was mostly constituted of monoterpenes, such as α-pinene or γ-terpinene, and especially of linalool, up to 81%. These results are consistent with previously published data [60,61]. Clary sage EO mostly consisted of oxygenated monoterpenes, up to 85% of the EO composition, such as linalool, β-myrcene, α-copaene, or β-caryophyllene. Linalyl acetate and linalool, both monoterpenes, are in fact the EO major compounds, as previously reported [20,26,41,48]. Moreover, the rather elevated amount of germacrene-D obtained in our experimental conditions corresponds to a previously described chemotype, rich in that specific compound [26,48,63].

However, even though the EO chemical composition for a same plant and plant part was highly similar between the tested experimental conditions (unpolluted or polluted soils), the relative abundances of several chemical compounds, such as linalool, linalyl acetate, or several aliphatic aldehydes, were found modified. Attention should be drawn to the influence of the environmental parameters on EO composition, such as geographic location, climate, soil conditions, along with cultivation practices [11,48,60]. Notably, the presence of elevated amounts of TE in soil has been shown to result in lower EO yields [43,44,64], or in altered EO chemical compositions, in response to the TE-induced stresses [37,39,65]. It is suspected that in response to TE exposure, inhibition or an activation of several key enzymes—involved in the biosynthesis pathways—could result in a modification of the plant secondary metabolism, and hence lead to either a reduction, or to an increase of specific secondary metabolites, respectively [43,64,65], which could explain the obtained differences. However, the variability among the experimental conditions was rather low, and the quality of the three different EO was faithful according to the chemotypes reported in the literature body [26,28,48,60,61], while the EO yields were in a related publication found unaffected by the soil pollution [34]. Furthermore, as previously highlighted [42,44], the response to TE exposure seems to vary greatly among aromatic plant species. In that regard, the chemical composition from clary sage inflorescence seemed to be less affected than the EO from coriander by the environmental conditions, and in particular by the presence of TE in soil. Aromatic plants from the genus *Salvia* (Lamiaceae) in particular, were in fact described as being able to tolerate elevated TE amounts, and to consistently grow in such conditions, displaying unaffected EO compositions [40,42], corroborating the obtained results. Finally, from a wider perspective, the variability recorded in terms of EO composition could be attributed not only to the presence of TE in soil, but also to the geographical location, and to the soil conditions [48,60]. In previously published data [34,45], it was indeed highlighted that the soil physico-chemical parameters were slightly different between the two experimental sites, which could explain that the two aromatic plants grown in situ displayed slightly different EO chemical compositions, even though the plant maturity stages at harvest were identical, and that TE in soil did not hinder plant growth [34].

4.2. EO Biological Activities towards a Potential Application in Crop Protection

EO from both coriander and clary sage were previously investigated for their antifungal activity against a wide spectrum of fungal pathogens, but reports targeting plant pathogens in particular are scarce. Until now, positive results have been reported regarding the antifungal effects of EO from coriander seeds, and from Lamiaceae species, on the development of *Fusarium* spp. or various other fungal phytopathogens [4,60]. As highlighted in the majority of the previous studies, EO biological effects are often dependent upon the EO concentration [4]. In the same way, our data regarding antifungal activity, against both *F. culmorum* and *Z. tritici*, have shown that all the EO that were evaluated in this study inhibited fungal growth. Notably, the observed antifungal effect was defined as fungistatic rather than fungicidal, depicted by the revival of hyphae and mycelial growth, after transfer on a medium exempted from EO. This feature could be valuable in preventive applications as a means to control pre- and postharvest fungal diseases. The efficiency of the EO increased in the following order: clary sage EO < coriander seeds EO ≤ coriander aerial parts EO. Although the concentrations of the EO were up to 1000 times higher than those of the chemical marketed fungicides, which is commonly observed [66], the tested EO still displayed consistent antifungal activity. Furthermore, EO are known to exert lower harmful effects on non-target organisms, and on environment and human health [4,11]. Notably, they are known to possess a low persistence in soils, owing to their volatility [4,67–69], while the occurrence of resistance phenomena bound to the use of EO has not been reported so far. This feature could be bound to their action as multisite chemicals [4]. It should also be noted that, in the case of *Z. tritici*, all the tested EO were in the same range of efficiency as the positive control consisting in a marketed fungicidal product. This feature could be particularly interesting. Indeed, by displaying a substantial biocidal activity, combined with a limited toxicity towards non-target organisms, and a high volatility hence limiting environmental risks, the tested EO appear as promising candidates, when compared to conventional pesticides or even other biocontrol products.

In addition to antifungal properties, the tested EO revealed a significant anti-germinative effect, and herbicidal activity, on both lettuce and rye-grass. Our results depicted a promising activity of the EO, especially the one distilled from aerial parts of coriander, which displayed lower IC_{50} values than those from coriander seeds and clary sage, whatever the bioassay, and the target plant. In comparison with glyphosate, which is a systemic herbicide, and was evaluated as a positive control in this work, the tested EO displayed a consistent herbicidal activity. In previous studies, reported glyphosate IC_{50} varied from 15.3 mg L^{-1} [70] to 23 and 46.2 mg L^{-1} [71] regarding the inhibition of ryegrass growth, while the results reported on lettuce ranged from 8.9 mg L^{-1} [72] to 20 mg L^{-1} [73], which are comparable to those obtained in this study. In addition, the EO from aerial parts of coriander, reported as the most efficient in terms of in vitro herbicidal activity, exerted effects similar to those of glyphosate, and even higher on rye-grass. Since glyphosate is not homologated as an anti-germinative product, the use of EO to fulfil this purpose could be promising.

4.3. EO Potential Applications as Post-Harvest Pests Control Products—Insecticidical Properties

Secondary metabolites from plants are also recognized to play a role in plant–insect interactions, and as such have been widely investigated for their insecticidal properties [26,74]. Their quick degradation could also favor their use as fumigants [11,75].

In this study, the potent repellent and fumigant activities were examined against the adults of *E. kuehniella*, *B. tabaci* and *R. dominica*.

In response to EO exposure, *E. kuehniella* and *B. tabaci* displayed mortality percentages up to 50%, whatever the EO, while coriander seeds EO displayed a mortality rate up to 100% against *R. dominica*. Moreover, whatever the EO concentration, the three tested EO displayed a similar range of efficiency against *B. tabaci*—the EO, and in particular those from coriander, resulted in a substantial insect mortality, even at low concentrations. Regarding repellence, the three tested EO displayed similar efficiencies against *B. tabaci*, while the EO from clary sage and from the aerial parts of coriander displayed the highest repellence percentages against *R. dominica* and *E. kuehniella*, respectively. Moreover, the obtained RC_{50} values ranged between 2.61 and 3.80 µL cm^{-2}, and between 2.61 and 3.77 µL cm^{-2}, against *E. kuehniella*, and *B. tabaci*, respectively, while those obtained against *R. dominica* were significantly lower, varying from 0.07 to 0.16 µL cm^{-2}. Finally, in the current fumigant bioassays, the three tested EO showed similar ranges of toxicity against the three target insects, ranging from 2.2 to 5.2 µL L^{-1}.

The insecticidal activity of several plant extracts and EO has previously been reported in several studies [26,66,67,74,75]. Little work has however been done using coriander or clary sage EO against the insects that are targeted in the present study.

Against *B. tabaci*, diverse EO were previously evaluated, such as those from *Citrus aurantium* peels, *Citrus sinensis*, *Allium sativum*, *Agastache rugosa*, *Illicium verum*, *Chenopodium ambrosioides*, *Schizonepeta tenuifolia*, *Curcuma aeruginosa*, *Syzygium aromaticum* or *Valeriana officinalis* [12,76,77]. Among all the tested EO, the strongest fumigant activities were obtained with the EO from *A. sativum*, *C. aurantium* and *A. rugosa*, with respective LC_{50} values of 0.11 µg L^{-1}, 3.97 and 5.8 µL L^{-1} and 7.08 µg L^{-1} [12,76,77]. In contrast, some EO did not result in any fumigant toxicity at the tested concentrations, such as those from *S. tenuifolia*, *C. aeruginosa* or *V. officinalis* [76]. Whatever the bioassay, the three EO tested in the current study displayed a fumigant activity, and similar efficiencies against *B. tabaci*. Moreover, in comparison with the body of literature, the obtained LC_{50} values would put them among the most efficient EO reported so far against *B. tabaci*.

Concerning *E. kuehniella*, EO from *Ocimum basilicum*, *Mentha pulegium* or *Ruta graveolens* previously displayed LC_{50} values ranging from 0.3 to 1.02 µL L^{-1} [11], while the one from *Pistacia lentiscus* was about 40.2 µL L^{-1} [11,78]. Essential oils from *Eucalyptus astringens*, *Eucalyptus leucoxylon*, *Eucalyptus lehmannii*, *Eucalyptus rudis*, *Eucalyptus camaldulensis*, and *Laurus nobilis* were also effective against *E. kuehniella*, since the related LC_{50} values ranged between 20.5 and 33.8 µL L^{-1} [11,79,80]. Thus, the results obtained during the present investigation suggest good potential for the three tested EO to be used as both fumigant and repellent products.

Coriander seeds EO, as well as its isolated major compounds, were previously evaluated against *R. dominica* [81]. A high mortality rate after 24 h of exposure (up to 100%) was observed, using a dose of 1 µL/15 mL of EO, which corroborates the high mortality percentages obtained with the EO distilled from seeds of coriander in the current experimental conditions. The EO from seeds of coriander has also been evaluated against several stored products pests, such as *Tribolium castaneum*, *Lasioderma serricorne* and *Sitophilus oryzae* [82], while the one distilled from coriander aerial parts was investigated for its effects against *T. castaneum* [83]. High inhibition of *T. castaneum* early development stages was observed [83], along with a significant fumigant toxicity reflected in LC_{50} values of 276.3, 5.3, and 145.5 µL L^{-1} of air, against *T. castaneum*, *L. serricorne* and *S. oryzae*, respectively [82]. The LC_{50} values acquired for the EO of clary sage, and coriander seeds and aerial parts, are this way within the same range of efficiency, and among the most efficient ones.

Against *R. dominica*, formulated aqueous extracts of clary sage were also previously reported for their toxic effects, with mortality rates above 95%, at the highest tested concentration [84]. These mortality rates, higher than those obtained in the present study, highlight the importance of an adequate EO formulation, so as to improve the EO biological effects as well as their persistence in time, often brought forward as limited [84–86].

4.4. Essential Oil Encapsulation in β-CD

To address this issue, EO encapsulation in cyclic oligosaccharides, such as β-CD, could help preventing EO oxidation, thermal degradation and quick evaporation, and allow a controlled-release of the EO and of their major compounds [87–90]. Interestingly, the EO studied in this work were previously demonstrated to be efficiently complexed with β-CD, since retention percentages ranged from 63 to 80% [20]. These are within the same range as those commonly described for some other EO [20,86]. It hence suggests that CD can efficiently retain EO and further reduce their volatility. However, in our experimental conditions, EO complexation with β-CD did not result in a significant improvement of the investigated biological properties. In some specific cases, notably in the antifungal and herbicidal assays, it even resulted in a lower efficiency of the EO (negative effect), owing to their complexation with β-CD and consequently their reduced volatility and availability. From an agricultural perspective, encapsulation could nonetheless significantly raise the persistence of the EO's effects, given their efficient retention by β-CD, allowing a controlled release [89,91]. It could particularly be valuable towards a lengthening of the fungistatic effects in time, which could then be of great interest to legitimize their use as natural alternatives.

Overall, our results suggest that the presence of TE in soil did not alter the EO biological effects, whatever the assessed property. Whether they were evaluated for applications in crop protection or as post-harvest treatments, EO originating from the biomass cultivated on the polluted site mostly displayed similar efficacies as the one distilled from the unpolluted one.

4.5. Insights on the Relationships between EO Composition and Their Biological Effects

Mono- and sesquiterpenoids are commonly described as responsible for the EO biological activities, whether they are antimicrobial, herbicidal or insecticidal [12,28,81,92,93]. As such, linalool, camphor and geranyl acetate were highlighted as the active compounds of the EO distilled from seeds of coriander, in terms of fumigant toxicity against *R. dominica* [82]. Similarly, clary sage EO insecticidal activity could be bound to its high amount of linalyl acetate in particular, since the exclusion of that compound from the EO mixture resulted in a substantial decrease in terms of repellence (halved) against a mite species, *Tetranychus urticae* [31]. From a wider perspective, linalool which is present in all the three tested EO, but in different proportions, is often highlighted as one of the main factors responsible for the EO bioactivity [81,94]. However, the variation observed between the different tested EO, whatever the biological property, cannot be explained by the action of their major components only.

In fact, it has been repeatedly emphasized that EO's biological effects were rather the result of a synergism between their compounds, since the evaluation of the latter isolated or of the mixture purified from one of its compounds, resulted in lower activities [12,28,95–97]. Since EO could act as multisite chemicals, lowering the risk of resistance phenomena [4], a deeper knowledge of their action mechanisms, and of some of their compounds, alone or in combination, would be of great interest. Even though the biological properties of a wide number of EO against various pathogenic microorganisms and pests have been covered, the investigation of the action mechanisms remains indeed limited. Several main features have nonetheless been highlighted regarding antifungal activity, such as the inhibition of the fungal cell wall formation, the disruption of the cell membrane (through the inhibition of the ergosterol synthesis), the inhibition of the mitochondrial electron transport, the inhibition of the cellular division, the interference with RNA, DNA synthesis and/or protein synthesis, and the inhibition of efflux pumps [5,98,99]. In that regard, coriander EO was demonstrated as efficient against *Candida albicans*, by increasing membrane permeability through a binding interaction with a membrane ergosterol [28,100]. Insecticidal activity of the EO, which has also been thoroughly investigated, points towards a site of action in the insect nervous system [66]. Plants' EO, and especially terpenoids compounds in it, seem to exhibit their toxicity through an interaction with different putative receptors, namely acetylcholinesterase, nicotinic acetylcholine receptor, octopamine receptor, or gamma-aminobutyric acid receptor ion channel [66,101]. They could moreover target multiple sites simultaneously [66], and act as insect repellents [81]. Regarding EO phytotoxic effects, resulting in visible symptoms, they can notably be the result of mitosis inhibition, a decrease of cellular respiration, ion leakage, membrane depolarization, decrease of the chlorophyll content, oxidative damages or removal of the cuticular waxy layer [5,102–104]. In the case of cinnamon and Java citronella EO, or of their main compounds, which could act as efficient herbicides, it was, for instance, demonstrated that the plant plasma membrane could be one of the EO's cellular targets, owing to the amphiphilic nature of several compounds [102]. The authors concluded that the mentioned EO or compounds were susceptible to affect lipid organization and/or domain formation, especially in the case of monoterpenes, while phenylpropanoids are likely to interact with membrane receptors [102]. However, no comprehensive study has so far been carried out on the detailed herbicidal mechanisms [102], which could be a valuable addition to the field.

5. Conclusions

The growing number of studies related to EO biological effects tends to demonstrate their suitability for the development of natural products-based biopesticides [4,17], provided that EO stability issues are solved. Our results demonstrate that growing aromatic plant on TE-polluted surfaces—and distilling EO from the grown biomass—could be a relevant tool to engage the reclaiming of these marginal lands.

As a whole, the obtained results indicate that the three evaluated EO, from coriander (aerial parts and seeds) and clary sage (inflorescences), displayed faithful chemical compositions, despite the soil pollution by TE. They also were able to inhibit the growth of two major fungal phytopathogens, while also exerting anti-germinative and herbicidal effects, against both mono- and dicotyledon species. Notably, the EO distilled from aerial parts of coriander possessed a higher efficiency, whatever the tested biological activity. Interestingly, significant repellent and fumigant activities were also demonstrated against three major post-harvest pests, whatever the EO. As a result, these EO could be promising candidates for the development of new biopesticides. Nonetheless, if such in vitro assays may indicate the EO's potential towards applications in crop protection or as post-harvest decay control products, these effects need to be confirmed by further *in planta* or in vivo assays, so as to legitimate their use. Moreover, even though the encapsulation of the tested EO in β-CD did not result in any improvement of the biological properties, further assessments should be conducted to confirm the efficiency of the controlled release of EO in glasshouse or field conditions.

Furthermore, these EO could be tested in combination with conventional marketed products, as well as with other EO or biocontrol products, so as to reduce the amounts used, or investigate potential synergistic effects.

Supplementary Materials: The following supporting information can be downloaded at: https://www.mdpi.com/article/10.3390/foods11030312/s1, Figure S1: Inhibitory effect of increasing EO concentrations from aerial parts of coriander, on the mycelial growth of *F. culmorum*, incubated for seven days, Figure S2: Antifungal activity of the three tested EO (aerial parts of coriander, a; seeds of coriander, b; and clary sage, c) against *Z. tritici*. Results for the in vitro microplate assay are displayed as optical densities—means from 4 values per well, Figure S3: Herbicidal activity of increasing EO concentrations from aerial parts of coriander, against *L. perenne* (a) and *L. sativa* (b), Table S1: RC_{50} values ($\mu L\ cm^{-2}$) for the three different tested EO, from the two experimental plots—after 24 h of exposure—against *E. kuehniella*, *B. tabaci* and *R. dominica* adults (n = 3).

Author Contributions: Conceptualization, methodology, validation, R.R., J.F., F.L. and A.L.-H.S.; writing—original draft preparation, R.R.; writing—review and editing, R.R., J.F., F.L., J.M.B.J., A.S. and A.L.-H.S.; supervision A.L.-H.S.; project administration, A.L.-H.S.; funding acquisition, A.L.-H.S. and J.F. All authors have read and agreed to the published version of the manuscript.

Funding: This work was supported by l'Agence De l'Environnement et de la Maîtrise de l'Energie (ADEME, Angers, France) in the framework of PhytEO and DEPHYTOP projects. This work has also been carried out in the framework of the Alibiotech project which is financed by the European Union, the French State and the French Region of Hauts-de-France.

Institutional Review Board Statement: Not applicable.

Informed Consent Statement: Not applicable.

Data Availability Statement: The original contributions presented in the study are publicly available.

Acknowledgments: The authors wish to thank the "Université du Littoral Côte d'Opale" and the "Pôle Métropolitain de la Côte d'Opale" for providing financial support for R. Raveau's Ph.D thesis. The authors also wish to acknowledge P. Ferrant from Ferrant PHE for his technical help, and EO distillation, as well as N. Facon for her technical help.

Conflicts of Interest: The authors declare no conflict of interest.

References

1. Camele, I.; Elshafie, H.S.; Caputo, L.; De Feo, V. Anti-quorum Sensing and Antimicrobial Effect of Mediterranean Plant Essential Oils Against Phytopathogenic Bacteria. *Front. Microbiol.* **2019**, *10*, 2619. [CrossRef] [PubMed]
2. Michel, J.; Abd Rani, N.Z.; Husain, K. A Review on the Potential Use of Medicinal Plants From *Asteraceae* and *Lamiaceae* Plant Family in Cardiovascular Diseases. *Front. Pharmacol.* **2020**, *11*, 1–26. [CrossRef] [PubMed]
3. Zaïri, A.; Nouir, S.; Zarrouk, A.; Haddad, H.; Khélifa, A.; Achour, L.; Tangy, F.; Chaouachi, M.; Trabelsi, M. Chemical composition, Fatty acids profile and Biological properties of *Thymus capitatus* (L.) Hoffmanns, essential Oil. *Sci. Rep.* **2019**, *9*, 1–8. [CrossRef]
4. De Clerck, C.; Dal Maso, S.; Parisi, O.; Dresen, F.; Zhiri, A.; Jijakli, M.H. Screening of Antifungal and Antibacterial Activity of 90 Commercial Essential Oils against 10 Pathogens of *Foods* **2020**, *9*, 1418. [CrossRef]
5. Raveau, R.; Fontaine, J.; Lounès-Hadj Sahraoui, A. Essential oils as potential alternative biocontrol products against plant pathogens and weeds: A review. *Foods* **2020**, *9*, 365. [CrossRef] [PubMed]
6. Barreca, S.; La Bella, S.; Maggio, A.; Licata, M.; Buscemi, S.; Leto, C.; Pace, A.; Tuttolomondo, T. Flavouring Extra-Virgin Olive Oil with Aromatic and Medicinal Plants Essential Oils Stabilizes Oleic Acid Composition during Photo-Oxidative Stress. *Agriculture* **2021**, *11*, 266. [CrossRef]
7. Mejri, S.; Siah, A.; Coutte, F.; Magnin-Robert, M.; Randoux, B.; Tisserant, B.; Krier, F.; Jacques, P.; Reignault, P.; Halama, P. Biocontrol of the wheat pathogen *Zymoseptoria tritici* using cyclic lipopeptides from *Bacillus subtilis*. *Environ. Sci. Pollut. Res.* **2018**, *25*, 29822–29833. [CrossRef] [PubMed]
8. Hellin, P.; King, R.; Urban, M.; Hammond-Kosack, K.E.; Legrève, A. The adaptation of *Fusarium culmorum* to DMI fungicides is mediated by major transcriptome modifications in response to azole fungicide, including the overexpression of a PDR transporter (FcABC1). *Front. Microbiol.* **2018**, *9*, 1–15. [CrossRef]
9. Ji, F.; He, D.; Olaniran, A.O.; Mokoena, M.P.; Xu, J.; Shi, J. Occurrence, toxicity, production and detection of *Fusarium mycotoxin*: A review. *Food Prod. Process. Nutr.* **2019**, *1*, 1–14. [CrossRef]
10. Khemira, S.; Jemaa, J.M.B.; Haouel, S.; Khouja, M.L. Repellent activity of essential oil of eucalyptus astringens against *rhyzopertha dominica* and *oryzaephilus surinamensis*. *Acta Hortic.* **2013**, *997*, 207–214. [CrossRef]
11. Chaaban, S.B.; Hamdi, S.H.; Mahjoubi, K.; Jemâa, J.M. Ben Composition and insecticidal activity of essential oil from *Ruta graveolens*, *Mentha pulegium* and *Ocimum basilicum* against *Ectomyelois ceratoniae* Zeller and *Ephestia kuehniella* Zeller (*Lepidoptera*: *Pyralidae*). *J. Plant Dis. Prot.* **2019**, *126*, 237–246. [CrossRef]
12. Zarrad, K.; Hamouda, A.B.; Chaieb, I.; Laarif, A.; Jemâa, J.M. Ben Chemical composition, fumigant and anti-acetylcholinesterase activity of the Tunisian Citrus *Aurantium* L. essential oils. *Ind. Crops Prod.* **2015**, *76*, 121–127. [CrossRef]
13. Yuan, L.; Wang, S.; Zhou, J.; Du, Y.; Zhang, Y.; Wang, J. Status of insecticide resistance and associated mutations in Q-biotype of whitefly, *Bemisia tabaci*, from eastern China. *Crop Prot.* **2012**, *31*, 67–71. [CrossRef]
14. Lundin, O.; Rundlöf, M.; Smith, H.G.; Fries, I.; Bommarco, R. Neonicotinoid insecticides and their impacts on bees: A systematic review of research approaches and identification of knowledge gaps. *PLoS ONE* **2015**, *10*, 1–20. [CrossRef]
15. Lamichhane, J.R.; Dachbrodt-Saaydeh, S.; Kudsk, P.; Messéan, A. Conventional Pesticides in Agriculture: Benefits Versus Risks. *Plant Dis.* **2016**, *100*, 10–24. [CrossRef]
16. Barzman, M.; Bàrberi, P.; Birch, A.N.E.; Boonekamp, P.; Dachbrodt-Saaydeh, S.; Graf, B.; Hommel, B.; Jensen, J.E.; Kiss, J.; Kudsk, P.; et al. Eight principles of integrated pest management. *Agron. Sustain. Dev.* **2015**, *35*, 1199–1215. [CrossRef]
17. Pavela, R.; Benelli, G. Essential Oils as Ecofriendly Biopesticides? Challenges and Constraints. *Trends Plant Sci.* **2016**, *21*, 1000–1007. [CrossRef]
18. Gossen, B.D.; McDonald, M.R. New technologies could enhance natural biological control and disease management and reduce reliance on synthetic pesticides. *Can. J. Plant Pathol.* **2019**, *42*, 30–40. [CrossRef]

19. El-Alam, I.; Raveau, R.; Fontaine, J.; Verdin, A.; Laruelle, F.; Fourmentin, S.; Chahine, R.; Makhlouf, H.; Lounès-Hadj Sahraoui, A. Antifungal and phytotoxic activities of essential oils: In vitro assays and their potential use in crop protection. *Agronomy* **2020**, *10*, 825. [CrossRef]
20. Raveau, R.; Fontaine, J.; Verdin, A.; Mistrulli, L.; Laruelle, F.; Fourmentin, S.; Lounès–Hadj Sahraoui, A. Chemical composition, antioxidant and anti-inflammatory activities of clary sage and coriander essential oils produced on polluted and amended soils-phytomanagement approach. *Molecules* **2021**, *26*, 5321. [CrossRef]
21. De-Montijo-Prieto, S.; del Razola, C. Spices: Composition, Antioxidant, and Antimicrobial Activities. *Biology* **2021**, *10*, 1091. [CrossRef] [PubMed]
22. Kfoury, M.; Auezova, L.; Greige-Gerges, H.; Fourmentin, S. Encapsulation in cyclodextrins to widen the applications of essential oils. *Environ. Chem. Lett.* **2019**, *17*, 129–143. [CrossRef]
23. Sim, L.Y.; Rani, N.Z.A.; Husain, K. *Lamiaceae*: An insight on their anti-allergic potential and its mechanisms of action. *Front. Pharmacol.* **2019**, *10*, 677. [CrossRef] [PubMed]
24. Singh, P.; Pandey, A.K. Prospective of essential oils of the *genus mentha* as biopesticides: A review. *Front. Plant Sci.* **2018**, *9*, 1–14. [CrossRef] [PubMed]
25. Blaskó, Á.; Gazdag, Z.; Gróf, P.; Máté, G.; Sárosi, S.; Krisch, J.; Vágvölgyi, C.; Makszin, L.; Pesti, M. Effects of clary sage oil and its main components, linalool and linalyl acetate, on the plasma membrane of *Candida albicans*: An in vivo EPR study. *Apoptosis* **2017**, *22*, 175–187. [CrossRef] [PubMed]
26. Aćimović, M.; Kiprovski, B.; Rat, M.; Sikora, V.; Popović, V.; Koren, A.; Brdar-Jokanović, M. *Salvia sclarea*: Chemical composition and biological activity. *J. Agron. Technol. Eng. Manag.* **2018**, *1*, 18–28.
27. Bhuiyan, M.N.I.; Begum, J.; Sultana, M. Chemical composition of leaf and seed essential oil of *Coriandrum sativum* L. from Bangladesh. *Bangladesh J. Pharmacol.* **2009**, *4*, 150–153. [CrossRef]
28. Freires, I.D.A.; Murata, R.M.; Furletti, V.F.; Sartoratto, A.; De Alencar, S.M.; Figueira, G.M.; Rodrigues, J.A.D.O.; Duarte, M.C.T.; Rosalen, P.L. *Coriandrum sativum* L. (Coriander) essential oil: Antifungal activity and mode of action on *Candida* spp., and molecular targets affected in human whole-genome expression. *PLoS ONE* **2014**, *9*, e99086. [CrossRef]
29. Karpiński, T.M. Essential oils of *lamiaceae* family plants as antifungals. *Biomolecules* **2020**, *10*, 103. [CrossRef]
30. Yoon, C.; Kang, S.H.; Jang, S.A.; Kim, Y.J.; Kim, G.H. Repellent Efficacy of Caraway and Grapefruit Oils for *Sitophilus oryzae* (*Coleoptera: Curculionidae*). *J. Asia. Pac. Entomol.* **2007**, *10*, 263–267. [CrossRef]
31. Yoon, J.; Tak, J.-H. Toxicity and Repellent Activity of Plant Essential Oils and Their Blending Effects Against Two Spotted Spider Mites, *Tetranychus urticae* Koch. *Korean Soc. Appl. Entomol.* **2018**, *57*, 199–207.
32. Ramsha, A.; Saleem, K.A.; Saba, B. Repellent Activity of Certain Plant Extracts (Clove, Coriander, Neem and Mint) Against Red Flour Beetle. *Am. Sci. Res. J. Eng. Technol. Sci.* **2019**, *55*, 83–91.
33. Pandey, J.; Verma, R.K.; Singh, S. Suitability of aromatic plants for phytoremediation of heavy metal contaminated areas: A review. *Int. J. Phytoremediation* **2019**, *21*, 405–418. [CrossRef]
34. Raveau, R.; Fontaine, J.; Bert, V.; Perlein, A.; Tisserant, B.; Ferrant, P.; Lounès-Hadj Sahraoui, A. In situ cultivation of aromatic plant species for the phytomanagement of an aged-trace element polluted soil: Plant biomass improvement options and techno-economic assessment of the essential oil production channel. *Sci. Total Environ.* **2021**, *789*, 147944. [CrossRef] [PubMed]
35. He, Z.; Shentu, J.; Yang, X.; Baligar, V.C.; Zhang, T.; Stoffella, P.J. Heavy Metal Contamination of Soils: Sources, Indicators, and Assessment. *J. Environ. Indic.* **2015**, *9*, 17–18.
36. Burges, A.; Alkorta, I.; Epelde, L.; Garbisu, C. From phytoremediation of soil contaminants to phytomanagement of ecosystem services in metal contaminated sites. *Int. J. Phytoremediat.* **2018**, *20*, 384–397. [CrossRef]
37. Sá, R.A.; Sá, R.A.; Alberton, O.; Gazim, Z.C.; Laverde, A.; Caetano, J.; Amorin, A.C.; Dragunski, D.C. Phytoaccumulation and effect of lead on yield and chemical composition of *Mentha crispa* essential oil. *Desalin. Water Treat.* **2015**, *53*, 3007–3017. [CrossRef]
38. Evangelou, M.W.H.; Papazoglou, E.G.; Robinson, B.H.; Schulin, R. Phytomanagement: Phytoremediation and the Production of Biomass for Economic Revenue on Contaminated Land. In *Phytoremediation: Management of Environmental Contaminants, Volume 1*; Springer: Berlin/Heidelberg, Germany, 2015; ISBN 9783319103952.
39. Asgari Lajayer, B.; Ghorbanpour, M.; Nikabadi, S. Heavy metals in contaminated environment: Destiny of secondary metabolite biosynthesis, oxidative status and phytoextraction in medicinal plants. *Ecotoxicol. Environ. Saf.* **2017**, *145*, 377–390. [CrossRef]
40. Angelova, V.R.; Ivanova, R.; Todorov, G.M.; Ivanov, K.I. Potential of *Salvia sclarea* L. for Phytoremediation of Soils Contaminated with Heavy Metals. *Int. J. Agric. Biosyst. Eng.* **2016**, *10*, 780–790. [CrossRef]
41. Kumar, R.; Sharma, S.; Pathania, V. Effect of shading and plant density on growth, yield and oil composition of clary sage (*Salvia sclarea* L.) in north western Himalaya. *J. Essent. Oil Res.* **2013**, *25*, 23–32. [CrossRef]
42. Biswas, S.; Koul, M.; Bhatnagar, A.K. Effect of salt, drought and metal stress on essential oil yield and quality in plants. *Nat. Prod. Commun.* **2011**, *6*, 1559–1564. [CrossRef] [PubMed]
43. Jezler, C.N.; De Almeida, A.F.; De Jesus, R.M.; De Oliveira, R.A. Pb and Cd on growth, leaf ultrastructure and essential oil yield mint (*Mentha arvensis* L.). *Crop Prod.* **2015**, *45*, 392–398. [CrossRef]
44. Prasad, A.; Chand, S.; Kumar, S.; Chattopadhyay, A.; Patra, D.D. Heavy Metals Affect Yield, Essential Oil Compound, and Rhizosphere Microflora of Vetiver (*Vetiveria zizanioides* Linn. nash) Grass. *Commun. Soil Sci. Plant Anal.* **2014**, *45*, 1511–1522. [CrossRef]

45. Raveau, R.; Fontaine, J.; Hijri, M.; Lounès-Hadj Sahraoui, A. The Aromatic Plant Clary Sage Shaped Bacterial Communities in the Roots and in the Trace Element-Contaminated Soil More Than Mycorrhizal Inoculation–A Two-Year Monitoring Field Trial. *Front. Microbiol.* **2020**, *11*, 1–18. [CrossRef] [PubMed]
46. Tibaldi, G.; Fontana, E.; Nicola, S. Cultivation practices do not change the *Salvia sclarea* L. essential oil but drying process does. *J. Food Agric. Environ.* **2010**, *8*, 790–794.
47. De Figueiredo, R.O.; Marques, M.O.M.; Nakagawa, J.; Ming, L.C. Composition of coriander essential oil from Brazil. *Acta Hortic.* **2004**, *629*, 135–137. [CrossRef]
48. Zutic, I.; Nitzan, N.; Chaimovitsh, D.; Schechter, A.; Dudai, N. Geographical location is a key component to effective breeding of clary sage (*Salvia sclarea*) for essential oil composition. *Isr. J. Plant Sci.* **2016**, *63*, 134–141. [CrossRef]
49. Siah, A.; Deweer, C.; Duyme, F.; Sanssené, J.; Durand, R.; Halama, P.; Reignault, P. Correlation of in planta endo-beta-1,4-xylanase activity with the necrotrophic phase of the hemibiotrophic fungus *Mycosphaerella graminicola*. *Plant Pathol.* **2010**, *59*, 661–670. [CrossRef]
50. Znini, M.; Cristofari, G.; Majidi, L.; El Harrak, A.; Paolini, J.; Costa, J. In vitro antifungal activity and chemical composition of *Warionia saharae* essential oil against 3 apple phytopathogenic fungi. *Food Sci. Biotechnol.* **2013**, *22*, 113–119. [CrossRef]
51. Sahmer, K.; Deweer, C.; Santorufo, L.; Louvel, B.; Douay, F. Utilisation d'une Régression non Linéaire pour des Applications Microbiologiques. In Modal Seminar. 2015. Available online: https://modal.lille.inria.fr/wikimodal/lib/exe/fetch.php?media=modal_sahmer_20150113.pdf (accessed on 9 September 2021).
52. Park, J.; Yoon, J.H.; Depuydt, S.; Oh, J.W.; Jo, Y.M.; Kim, K.; Brown, M.T.; Han, T. The sensitivity of an hydroponic lettuce root elongation bioassay to metals, phenol and wastewaters. *Ecotoxicol. Environ. Saf.* **2016**, *126*, 147–153. [CrossRef]
53. Judd, L.A.; Jackson, B.E.; Fonteno, W.C. Advancements in root growth measurement technologies and observation capabilities for container-grown plants. *Plants* **2015**, *4*, 369–392. [CrossRef] [PubMed]
54. Lyu, J.; Park, J.; Kumar Pandey, L.; Choi, S.; Lee, H.; De Saeger, J.; Depuydt, S.; Han, T. Testing the toxicity of metals, phenol, effluents, and receiving waters by root elongation in *Lactuca sativa* L. *Ecotoxicol. Environ. Saf.* **2018**, *149*, 225–232. [CrossRef] [PubMed]
55. Paul, A.L.; Daugherty, C.J.; Bihn, E.A.; Chapman, D.K.; Norwood, K.L.L.; Ferl, R.J. Transgene expression patterns indicate that spaceflight affects stress signal perception and transduction in Arabidopsis. *Plant Physiol.* **2001**, *126*, 613–621. [CrossRef] [PubMed]
56. Yazdanbakhsh, N.; Fisahn, J. Analysis of Arabidopsis thaliana root growth kinetics with high temporal and spatial resolution. *Ann. Bot.* **2010**, *105*, 783–791. [CrossRef] [PubMed]
57. Jilani, G.; Saxena, R.C. Repellent and Feeding Deterrent Effects of Turmeric Oil, Sweetflag Oil, Neem Oil, and a Neem-Based Insecticide Against Lesser Grain Borer (*Coleoptera*: *Bostrychidae*). *J. Econ. Entomol.* **1990**, *83*, 629–634. [CrossRef]
58. Nerio, L.S.; Olivero-Verbel, J.; Stashenko, E.E. Repellent activity of essential oils from seven aromatic plants grown in Colombia against *Sitophilus zeamais* Motschulsky (*Coleoptera*). *J. Stored Prod. Res.* **2009**, *45*, 212–214. [CrossRef]
59. R Core Team. *R: A Language and Environment for Statistical Computing*; R Foundation for Statistical Computing: Vienna, Austria, 2019.
60. Sumalan, R.M.; Alexa, E.; Popescu, I.; Negrea, M.; Radulov, I.; Obistioiu, D.; Cocan, I. Exploring ecological alternatives for crop protection using *Coriandrum sativum* essential oil. *Molecules* **2019**, *24*, 2040. [CrossRef]
61. Satyal, P.; Setzer, W.N. Chemical Compositions of Commercial Essential Oils From *Coriandrum sativum* Fruits and Aerial Parts. *Nat. Prod. Commun.* **2020**, *15*, 1934578X20933067. [CrossRef]
62. Pavela, R.; Žabka, M.; Bednář, J.; Tříska, J.; Vrchotová, N. New knowledge for yield, composition and insecticidal activity of essential oils obtained from the aerial parts or seeds of fennel (*Foeniculum vulgare* Mill.). *Ind. Crops Prod.* **2016**, *83*, 275–282. [CrossRef]
63. Sharopov, F.S.; Setzer, W.N. The essential oil of *salvia sclarea* L. from Tajikistan. *Rec. Nat. Prod.* **2012**, *6*, 75–79.
64. Amirmoradi, S.; Moghaddam, P.R.; Koocheki, A.; Danesh, S.; Fotovat, A. Effect of *Cadmium* and Lead on Quantitative and Essential Oil Traits of Peppermint (*Mentha piperita* L.). *Not. Sci. Biol.* **2012**, *4*, 101–109. [CrossRef]
65. Gautam, M.; Agrawal, M. Influence of metals on essential oil content and composition of lemongrass (*Cymbopogon citratus* (D.C.) Stapf.) grown under different levels of red mud in sewage sludge amended soil. *Chemosphere* **2017**, *175*, 315–322. [CrossRef] [PubMed]
66. Isman, M.B. Bioinsecticides based on plant essential oils: A short overview. *Z. Naturforsch.-Sect. C J. Biosci.* **2020**, *75*, 179–182. [CrossRef] [PubMed]
67. Isman, M.B.; Machial, C.M. Chapter 2 Pesticides based on plant essential oils: From traditional practice to commercialization. *Adv. Phytomed.* **2006**, *3*, 29–44. [CrossRef]
68. Regnault-Roger, C.; Vincent, C.; Arnason, J.T. Essential oils in insect control: Low-risk products in a high-stakes world. *Annu. Rev. Entomol.* **2012**, *57*, 405–424. [CrossRef]
69. Seixas, P.T.L.; Demuner, A.J.; Alvarenga, E.S.; Barbosa, L.C.A.; Marques, A.; Farias, E.D.S.; Picanço, M.C. Bioactivity of essential oils from artemisia against *diaphania hyalinata* and its selectivity to beneficial insects. *Sci. Agric.* **2018**, *75*, 519–525. [CrossRef]
70. Ghanizadeh, H.; Harrington, K.C.; James, T.K.; Woolley, D.J. Quick tests for detecting glyphosate-resistant Italian and perennial ryegrass. *N. Z. J. Agric. Res.* **2015**, *58*, 108–120. [CrossRef]

71. Martin, M.L.; Ronco, A.E. Effect of mixtures of pesticides used in the direct seeding technique on nontarget plant seeds. *Bull. Environ. Contam. Toxicol.* **2006**, *77*, 228–236. [CrossRef]
72. Ronco, A.E.; Carriquiriborde, P.; Natale, G.S.; Martin, M.L.; Mugni, H.; Bonetto, C. Integrated approach for the assessment of biotech soybean pesticides impact on low order stream ecosystems of the pampasic region. In *Ecosystem Ecology Research Trends*; Nova Science Publishers, Inc.: New York, NY, USA, 2008; ISBN 9781604561838.
73. Kuang, Y.; Yang, S.X.; Sampietro, D.A.; Zhang, X.F.; Tan, J.; Gao, Q.X.; Liu, H.W.; Ni, Q.X.; Zhang, Y.Z. Phytotoxicity of leaf constituents from bamboo (*Shibataea chinensis nakai*) on germination and seedling growth of lettuce and cucumber. *Allelopath. J.* **2017**, *40*, 133–142. [CrossRef]
74. Benelli, G.; Pavela, R.; Maggi, F.; Petrelli, R.; Nicoletti, M. Commentary: Making Green Pesticides Greener? The Potential of Plant Products for Nanosynthesis and Pest Control. *J. Clust. Sci.* **2017**, *28*, 3–10. [CrossRef]
75. Yang, Y.; Isman, M.B.; Tak, J.H. Insecticidal activity of 28 essential oils and a commercial product containing *cinnamomum cassia* bark essential oil against *sitophilus zeamais* Motschulsky. *Insects* **2020**, *11*, 474. [CrossRef] [PubMed]
76. Liu, X.C.; Hu, J.F.; Zhou, L.; Liu, Z.L. Evaluation of fumigant toxicity of essential oils of Chinese medicinal herbs against Bemisia tabaci (Gennadius)(Hemiptera: Aleyrodidae). *J. Entomol. Zool. Stud.* **2014**, *2*, 164–169.
77. de Carvalho Ribeiro, N.; Gomes da Camara, C.A.; de Souza Born, F.; Abreu de Siqueira, H.A. Insecticidal Activity Against *Bemisia tabaci* Biotype B of Peel Essential Oil of *Citrus sinensis* var. pear and *Citrus aurantium* Cultivated in Northeast Brazil. *Nat. Prod. Commun.* **2010**, *5*, 1819–1822. [CrossRef]
78. Bachrouch, O.; Mediouni-Ben Jemâa, J.; Wissem, A.W.; Talou, T.; Marzouk, B.; Abderraba, M. Composition and insecticidal activity of essential oil from *Pistacia lentiscus* L. against *Ectomyelois ceratoniae* Zeller and *Ephestia kuehniella* Zeller (Lepidoptera: Pyralidae). *J. Stored Prod. Res.* **2010**, *46*, 242–247. [CrossRef]
79. Mediouni Ben Jemâa, J. Essential Oil as a Source of Bioactive Constituents for the Control of Insect Pests of Economic Importance in Tunisia. *Med. Aromat. Plants* **2014**, *03*, 158. [CrossRef]
80. Mediouni Ben Jemâa, J.; Tersim, N.; Boushih, E.; Taleb-toudert, K.; Larbi Khouja, M. Fumigant Control of the Mediterranean Flour Moth *Ephestia kuehniella* with the Noble Laurel *Laurus nobilis* Essential Oils. *Tunis. J. Plant Prot.* **2013**, *8*, 33–44.
81. López, M.D.; Jordán, M.J.; Pascual-Villalobos, M.J. Toxic compounds in essential oils of coriander, caraway and basil active against stored rice pests. *J. Stored Prod. Res.* **2008**, *44*, 273–278. [CrossRef]
82. Sriti Eljazi, J.; Bachrouch, O.; Salem, N.; Msaada, K.; Aouini, J.; Hammami, M.; Boushih, E.; Abderraba, M.; Limam, F.; Mediouni Ben Jemaa, J. Chemical composition and insecticidal activity of essential oil from coriander fruit against *Tribolium castaenum*, *Sitophilus oryzae*, and *Lasioderma serricorne*. *Int. J. Food Prop.* **2018**, *20*, S2833–S2845. [CrossRef]
83. Islam, M.S.; Hasan, M.M.; Xiong, W.; Zhang, S.C.; Lei, C.L. Fumigant and repellent activities of essential oil from *Coriandrum sativum* (L.) (*Apiaceae*) against red flour beetle *Tribolium castaneum* (Herbst) (*Coleoptera: Tenebrionidae*). *J. Pest Sci.* **2009**, *82*, 171–177. [CrossRef]
84. Sucur, J.; Gvozdenac, S.; Anackov, G.; Malencic, D.; Prvulovic, D. Allelopathic effects of *Clinopodium menthifolium* and *Salvia sclarea* aqueous extracts. *Zb. Matice Srp. Za Prir. Nauk.* **2016**, *131*, 177–188. [CrossRef]
85. Kfoury, M.; Hădărugă, N.G.; Hădărugă, D.I.; Fourmentin, S. Cyclodextrins as Encapsulation Material for Flavors and Aroma. *Encapsulations* **2016**, *2*, 127–192.
86. Kfoury, M.; Auezova, L.; Greige-Gerges, H.; Fourmentin, S. Promising applications of cyclodextrins in food: Improvement of essential oils retention, controlled release and antiradical activity. *Carbohydr. Polym.* **2015**, *131*, 264–272. [CrossRef] [PubMed]
87. Kfoury, M.; Auezova, L.; Fourmentin, S.; Greige-Gerges, H. Investigation of monoterpenes complexation with hydroxypropyl-β-cyclodextrin. *J. Incl. Phenom. Macrocycl. Chem.* **2014**, *80*, 51–60. [CrossRef]
88. Kfoury, M.; Landy, D.; Fourmentin, S. Contribution of headspace to the analysis of cyclodextrin inclusion complexes. *J. Incl. Phenom. Macrocycl. Chem.* **2019**, *93*, 19–32. [CrossRef]
89. Abarca, R.L.; Rodríguez, F.J.; Guarda, A.; Galotto, M.J.; Bruna, J.E. Characterization of beta-cyclodextrin inclusion complexes containing an essential oil component. *Food Chem.* **2016**, *196*, 968–975. [CrossRef]
90. Campos, E.V.R.; Proença, P.L.F.; Oliveira, J.L.; Pereira, A.E.S.; De Morais Ribeiro, L.N.; Fernandes, F.O.; Gonçalves, K.C.; Polanczyk, R.A.; Pasquoto-Stigliani, Y.; Lima, R.; et al. Carvacrol and linalool co-loaded in β-cyclodextrin-grafted chitosan nanoparticles as sustainable biopesticide aiming pest control. *Sci. Rep.* **2018**, *8*, 1–14. [CrossRef]
91. Ciobanu, A.; Mallard, I.; Landy, D.; Brabie, G.; Nistor, D.; Fourmentin, S. Retention of aroma compounds from *Mentha piperita* essential oil by cyclodextrins and crosslinked cyclodextrin polymers. *Food Chem.* **2013**, *138*, 291–297. [CrossRef]
92. Choi, H.-J.; Sowndhararajan, K.; Cho, N.-G.; Hwang, K.-H.; Koo, S.-J.; Kim, S. Evaluation of Herbicidal Potential of Essential Oils and their Components under In vitro and Greenhouse Experiments. *Weed Turfgrass Sci.* **2015**, *4*, 321–329. [CrossRef]
93. Matasyoh, J.C.; Maiyo, Z.C.; Ngure, R.M.; Chepkorir, R. Chemical composition and antimicrobial activity of the essential oil of *Coriandrum sativum*. *Food Chem.* **2009**, *113*, 526–529. [CrossRef]
94. Duarte, A.; Luís, Â.; Oleastro, M.; Domingues, F.C. Antioxidant properties of coriander essential oil and linalool and their potential to control *Campylobacter* spp. *Food Control* **2016**, *61*, 115–122. [CrossRef]
95. Bakkali, F.; Averbeck, S.; Averbeck, D.; Idaomar, M. Biological effects of essential oils–A review. *Food Chem. Toxicol.* **2008**, *46*, 446–475. [CrossRef] [PubMed]
96. Hu, J.; Wang, W.; Dai, J.; Zhu, L. Chemical composition and biological activity against *Tribolium castaneum* (Coleoptera: Tenebrionidae) of *Artemisia brachyloba* essential oil. *Ind. Crops Prod.* **2019**, *128*, 29–37. [CrossRef]

97. Bougherra, H.H.; Bedini, S.; Flamini, G.; Cosci, F.; Belhamel, K.; Conti, B. *Pistacia lentiscus* essential oil has repellent effect against three major insect pests of pasta. *Ind. Crops Prod.* **2015**, *63*, 249–255. [CrossRef]
98. Lagrouh, F.; Dakka, N.; Bakri, Y. The antifungal activity of Moroccan plants and the mechanism of action of secondary metabolites from plants. *J. Mycol. Med.* **2017**, *27*, 303–311. [CrossRef] [PubMed]
99. Nazzaro, F.; Fratianni, F.; Coppola, R.; De Feo, V. Essential oils and antifungal activity. *Pharmaceuticals* **2017**, *10*, 86. [CrossRef]
100. Silva, F.; Ferreira, S.; Queiroz, J.A.; Domingues, F.C. Coriander (*Coriandrum sativum* L.) essential oil: Its antibacterial activity and mode of action evaluated by flow cytometry. *J. Med. Microbiol.* **2011**, *60*, 1479–1486. [CrossRef]
101. Kim, S.; Yoon, J.; Tak, J.H. Synergistic mechanism of insecticidal activity in basil and mandarin essential oils against the tobacco cutworm. *J. Pest Sci.* **2021**, *94*, 1119–1131. [CrossRef]
102. Lins, L.; Dal Maso, S.; Foncoux, B.; Kamili, A.; Laurin, Y.; Genva, M.; Jijakli, M.H.; De Clerck, C.; Fauconnier, M.L.; Deleu, M. Insights into the relationships between herbicide activities, molecular structure and membrane interaction of cinnamon and citronella essential oils components. *Int. J. Mol. Sci.* **2019**, *20*, 4007. [CrossRef] [PubMed]
103. Maffei, M.; Camusso, W.; Sacco, S. Effect of *Mentha* x *piperita* essential oil and monoterpenes on cucumber root membrane potential. *Phytochemistry* **2001**, *58*, 703–707. [CrossRef]
104. Zunino, M.P.; Zygadlo, J.A. Effect of monoterpenes on lipid oxidation in maize. *Planta* **2004**, *219*, 303–309. [CrossRef]

Article

Global Proteomic Analysis of *Listeria monocytogenes'* Response to Linalool

Zhipeng Gao [1,*], Weiming Zhong [1], Ting Liu [2], Tianyu Zhao [1] and Jiajing Guo [2,3,*]

[1] Hunan Engineering Technology Research Center of Featured Aquatic Resources Utilization, College of Animal Science and Technology, Hunan Agricultural University, Changsha 410128, China; zhongweiming2021@163.com (W.Z.); z1583025825@163.com (T.Z.)
[2] Hunan Agriculture Product Processing Institute, Hunan Academy of Agricultural Sciences, Changsha 410125, China; ltchangsha98@163.com
[3] Key Laboratory of Agro-Products Processing, Ministry of Agriculture and Rural Affairs of P. R. China, Institute of Food Science and Technology CAAS, Beijing 100193, China
* Correspondence: gaozhipeng627@163.com (Z.G.); guojiajing1986@163.com (J.G.)

Citation: Gao, Z.; Zhong, W.; Liu, T.; Zhao, T.; Guo, J. Global Proteomic Analysis of *Listeria monocytogenes'* Response to Linalool. *Foods* **2021**, *10*, 2449. https://doi.org/10.3390/foods10102449

Academic Editors: Lisa Pilkington and Siew-Young Quek

Received: 30 August 2021
Accepted: 13 October 2021
Published: 14 October 2021

Publisher's Note: MDPI stays neutral with regard to jurisdictional claims in published maps and institutional affiliations.

Copyright: © 2021 by the authors. Licensee MDPI, Basel, Switzerland. This article is an open access article distributed under the terms and conditions of the Creative Commons Attribution (CC BY) license (https:// creativecommons.org/licenses/by/ 4.0/).

Abstract: *Listeria monocytogenes* (LM) is one of the most serious foodborne pathogens. Listeriosis, the disease caused by LM infection, has drawn attention worldwide because of its high hospitalization and mortality rates. Linalool is a vital constituent found in many essential oils; our previous studies have proved that linalool exhibits strong anti-*Listeria* activity. In this study, iTRAQ-based quantitative proteomics analysis was performed to explore the response of LM exposed to linalool, and to unravel the mode of action and drug targets of linalool against LM. A total of 445 differentially expressed proteins (DEPs) were screened out, including 211 up-regulated and 234 down-regulated proteins which participated in different biological functions and pathways. Thirty-one significantly enriched gene ontology (GO) functional categories were obtained, including 12 categories in "Biological Process", 10 categories in "Cell Component", and 9 categories in "Molecular Function". Sixty significantly enriched biological pathways were classified, including 6 pathways in "Cell Process", 6 pathways in "Environmental Information Processing", 3 pathways in "Human Disease", 40 pathways in "Metabolism", and 2 pathways in "Organic System". GO and Kyoto Encyclopedia of Genes (KEGG) enrichment analysis together with flow cytometry data implied that cell membranes, cell walls, nucleoids, and ribosomes might be the targets of linalool against LM. Our study provides good evidence for the proteomic analysis of bacteria, especially LM, exposed to antibacterial agents. Further, those drug targets discovered by proteomic analysis can provide theoretical support for the development of new drugs against LM.

Keywords: linalool; *Listeria monocytogenes*; antimicrobial; proteomics

1. Introduction

The prevention and control of foodborne pathogens is always an urgent need for food safety and human health worldwide [1]. *Listeria monocytogenes* is listed as one of the most serious foodborne pathogens by the World Health Organization (WHO) [2]. It is widely distributed in nature and can survive in many extreme environments such as high salt, low temperature, low pH, and so on [3,4]. Listeriosis caused by LM infection is a serious food-borne zoonotic disease with high hospitalization and mortality rates [5–8]. The clinical manifestations mainly include meningitis, septicemia, and endocarditis. Pregnant women, newborns, the elderly, and people with weakened immunity are susceptible groups. Among them, pregnant women are more likely to suffer from listeriosis, and severe cases can even cause premature birth, stillbirth, and neurological diseases in the offspring [3,9]. Thus, the control and prevention of LM has become a crucial issue all over the world.

Chemical antimicrobial agents are usually used for the prevention of LM in the food industry. However, today more and more consumers are likely to pursue "green and natural" foods with few or no chemicals. Yet natural antimicrobial agents, especially essential oils or their ingredients, have become a popular research area [10–13]. Linalool (3,7-dimethyl-1,6-octadien-3-ol) is a vital constituent found in many essential oils with good antibacterial activity against different kinds of microorganisms. It is generally recognized as a safe (GRAS) food additive [14]. In our previous study, we demonstrated that linalool was a major component in citrus essential oils and exhibited strong antibacterial activity against LM—the ZOI, MIC, and MBC values were 39.58 ± 0.74 mm, 0.5% (v/v), and 1% (v/v), respectively—and it exhibited significant anti-biofilm activity by the dispersal and killing of cells in biofilm [15–17], but little is known about its mode of action against LM.

Recently, with the rapid development of omics-technology, many omics, such as genomics, transcriptomics, and proteomics, have been used as effective research tools for microbiology study [18–20]. Among them, proteomics is often used to investigate bacterial behavior under different environmental conditions in protein levels [13,21]. Research focused on the theme of "changes of microbial proteomics after drug treatment" has become a hot topic. Especially in "antibiotic treatment" fields, some research groups have used proteomics technology on the following research objects: erythromycin against *Streptococcus suis* [22], vancomycin against *Streptomyces coelicolor* [23], daptomycin against *Staphylococcus aureus* [24], bostrycin against *Mycobacterium tuberculosis* [25], emodin against *Haemophilus parasuis* [26], oxytetracycline against *Edwardsiella tarda* [27], etc.

Thus, in this study, we performed a global protein analysis by using iTRAQ (isobaric tags for relative and absolute quantitation)-based quantitative proteomic technology [28,29] to explore how LM responds to the treatment of linalool, and to unravel the mode of action and the drug targets of linalool against LM. We hope this continuous research will provide more of a theoretical basis for the prevention and control of LM.

2. Materials and Methods

2.1. Bacterial Strains

The LM (ATCC 19115) strain was obtained from Guangdong Microbiology Culture Center (GMCC, Guangdong, China) and stored at −80 °C.

2.2. Linalool

Linalool solution (95%) was purchased from Sigma-Aldrich (Sigma-Aldrich, Burlington, MA, USA). The antimicrobial activity of linalool was tested in our previous study, which showed the MIC value was 0.5% (v/v) [16].

2.3. Treatment of LM by Linalool

LM was grown in Brain Heart Infusion broth (BHI, Guangdong Huankai Microbial, Guangdong, China) at 37 °C with shaking overnight and transferred to fresh BHI broth at a ratio of 1:50. When the growth state of LM reached the logarithmic phase, linalool at a concentration of 4 × MIC was added to the bacterial solution and cultured at 37 °C with shaking for 8 h. After that, the cells were centrifuged (4000× g, 10 min) and washed three times with sterile PBS. Finally, the cells were collected for both flow cytometry and proteomics assay. For the proteomics assay, cells were frozen in liquid nitrogen for 3 h and then stored at −80 °C before protein extraction.

2.4. iTRAQ-Based Quantitative Proteomics Analysis

2.4.1. Protein Extraction, Digestion, and Labeling with iTRAQ Reagents

The treatment of LM cells was mentioned in Section 2.3. According to the manufacturer's protocol (from the Majorbio company, Shanghai, China), LM cells were resuspended with a lysis buffer (cocktail of 1% SDS and 8 M urea) in the ratio of 1:8, sonicated (Fielda-650D, Jiangsu TRON Intelligent Technology Co., Ltd., Jiangsu, China) for 4 min, and incubated on ice for 30 min. The lysates were centrifuged at 12,000× g at 4 °C for 15 min and the supernatants were collected. The concentration of protein was determined by bicinchoninic acid (BCA) assay. Protein digestion was carried out according to the standard procedure and the resulting peptide mixture was labeled by using 8-plex iTRAQ reagents according to the instructions (Applied Biosystems, MA, USA). For 8-plex labeling, each iTRAQ reagent was dissolved in 50 µL of isopropanol, added to the peptide mixture, and incubated at room temperature for 2 h. The samples were labeled as (HN12-1)-115, (HN12-2)-116, (HN12-3)-117, (PG45-1)-118, (PG45-2)-119, and (PG45-3)-121. Finally, all the samples were mixed together and vacuum dried before LC-MS/MS Analysis.

2.4.2. Chromatographic Separation and LC-MS/MS Analysis

Samples were re-suspended by loading buffer (ammonium hydroxide solution containing 2% acetonitrile, pH 10) and separated by high-pH reversed phase liquid chromatography (RPLC, Waters, Milford, MA, USA). The gradient elution was performed on a high pH RPLC column at 400 µL/min with the gradient increased for 66 min. Twenty fractions were collected from each sample. LC-MS/MS analysis was performed by a Q Exactive mass spectrometer (Thermo Fisher Scientific, Waltham, MA, USA) coupled with an Easy-nLC 1200 (Thermo Fisher Scientific, Waltham, MA, USA) in the data-dependent mode. Survey full-scan MS spectra were acquired at a mass resolution of 70 K, followed by 20 sequential high energy collisional dissociation LC-MS/MS scans at a resolution of 17.5 K. One micro-scan was recorded by using a dynamic exclusion of 18 s.

2.4.3. Proteomic Analysis

All the LC-MS/MS spectra were searched by using the Protein Discoverer Software (ProteomeDiscoverer™ Software 2.4, Thermo Fisher Scientific, Waltham, MA, USA) against the Mycoplasma database. The highest score for a given peptide mass was used to identify parent proteins. The parameters for protein searching were as follows: tryptic digestion with up to two missed cleavages, carbamidomethylation of cysteine as the fixed modification, and oxidation of methionine and protein N-terminal acetylation as variable modifications. Peptide spectral matches were validated based on q-values at a 1% false discovery rate.

2.5. Flow Cytometry Analysis

Flow cytometry analysis was carried out to investigate the effects of linalool on LM. After treatment as mentioned in Section 2.3, the bacterial cells were collected by centrifugation and adjusted to the concentration of 1×10^6 cfu/mL; then, different staining procedures were proceeded as follows [30,31].

2.5.1. Membrane Permeability

Thiazole orange (TO, Sigma-Aldrich, Burlington, MA, USA) and propidium iodide (PI, Sigma-Aldrich, Burlington, MA, USA) were used to evaluate the membrane permeability of the cells. For TO staining, 1 µL of TO solution was added to 1 mL bacterial suspensions (final concentration of TO: 10 µg/mL in DMSO), and then incubated at room temperature for 15 min. For PI staining, 1 µg PI was added to 1 mL bacterial suspensions (final concentration of PI: 1 µg/mL in PBS) and then incubated at 37 °C for 15 min.

2.5.2. Membrane Potential

Bis-1,3-dibutylbarbituric acid (BOX, Sigma-Aldrich, MA, USA) and PI were used to evaluate the membrane potential of cells. For BOX staining, 2.5 µg of BOX was added to 1 mL bacterial suspensions (final concentration of BOX: 2.5 µg/mL in PBS with 4 mM EDTA), and then incubated at 37 °C for 15 min. For PI staining, the procedure was the same as mentioned in Section 2.5.1.

2.5.3. Efflux Activity

Ethidium bromide (EB, Sigma-Aldrich, Burlington, MA, USA) was used to evaluate the efflux activity of cells. For EB staining, 10 µg of EB was added to 1 mL bacterial suspensions (final concentration of EB: 10 µg/mL in DMSO), and then incubated at 37 °C for 15 min.

2.5.4. Respiratory Activity

5-cyano-2,3-ditolyl tetrazolium chloride (CTC, Sigma-Aldrich, MA, USA) was used to evaluate the respiratory activity of cells. For CTC staining, 5 mM of CTC was added to 1 mL bacterial suspensions (final concentration of CTC: 5 mM in PBS with 1% (w/v) glucose), and then incubated at 37 °C with shaking (250 rpm) for 30 min.

After finishing these staining procedures, samples were washed with PBS three times, the concentration of the bacterial suspension was adjusted to about OD600 = 0.1, and the samples were placed on ice for flow cytometry analysis by a flow cytometer (BD Accuri C6 plus, BD, Franklin Lakes, NJ, USA).

2.6. Statistical Analysis

All the experiments were performed in triplicate. Statistical analysis was performed by GraphPad Prism 7.0 for t-tests. All asterisks indicate significant differences ($p < 0.05$).

3. Results

3.1. Quality Assessment of Proteomics Sequencing

As shown in Figure 1, a total of 26,880 peptides and 2102 proteins were identified by proteomics sequencing (Figure 1C). The lengths of the peptides were mainly between 6–20 amino acids, among which 7 was the most common (Figure 1A). The number of peptides that make up proteins was mainly concentrated between 1–21, among which 1–3 was the most common (Figure 1B). Among those identified proteins, the molecular weight was mainly concentrated between 11–60 kDa, among which 21–30 kDa was the most common (Figure 1D).

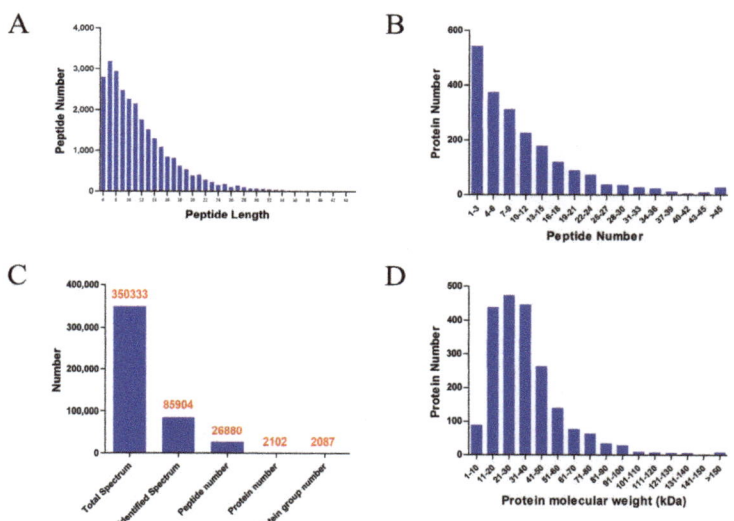

Figure 1. The quality assessments of proteomics sequencing. (**A**) Histogram of peptide length distribution. The abscissa represents the range of peptide length, and the ordinate represents the number of peptides of the corresponding length. (**B**) Histogram of peptide quantity distribution. The abscissa represents the range of the number of peptides covering the proteins, and the ordinate represents the number of the proteins. (**C**) Statistical histogram of different types of identified proteins.

(**D**) Histogram of protein molecular weight distribution. The abscissa represents the distribution range of protein molecular weight, and the ordinate represents the number of proteins corresponding to the molecular weight.

3.2. Functional Annotation and Analysis of Proteins

After quality assessment, GO and KEGG databases were used to annotate and analyze the functions of the identified proteins to explore their biological pathways and functions. As shown in Figure 2A, Biological Processes, Cell Components, and Molecular Functions (within which all the identified proteins were included) were obtained by GO classification annotation. Among "Molecular Functions", the top five were the catalytic activity, binding, transporter activity, transcription regulator activity, and structural molecule activity, and the number of proteins involved were 1192, 860, 164, 77, and 60, respectively. Among "Cell Components", cell part, membrane part, membrane, protein-containing complex, and organelle were the top five, and the number of proteins involved were 548, 424, 128, 102, and 56, respectively. Among "Biological Processes", metabolic process, cellular process, localization, biological regulation, and response to stimulus were the top five, and the number of proteins involved were 847, 795, 166, 161, and 79, respectively.

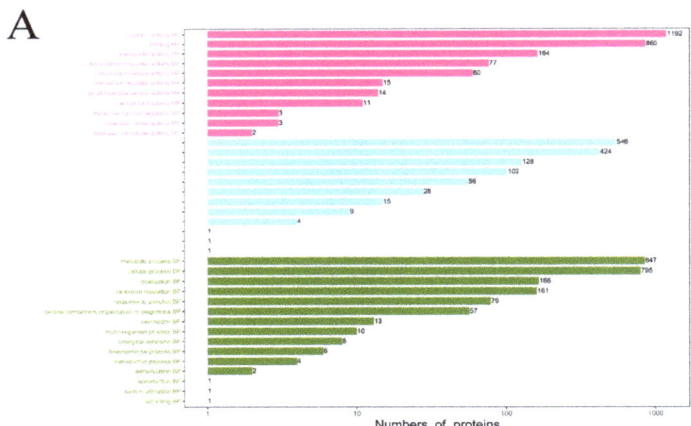

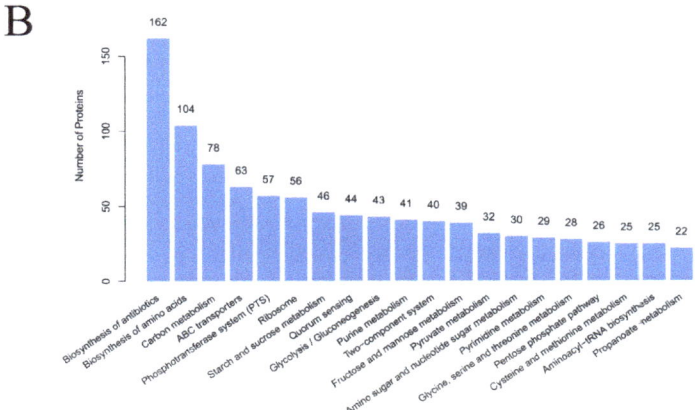

Figure 2. The annotation and analysis of all the identified proteins. (**A**) Annotation based on GO function classification. Each column represents a secondary classification. The ordinate represents

the secondary classification term of GO, and the capital letters in front of the term represent the following categories: BP, biological process; CC, cellular component; MF, molecular function. (**B**) The top 20 KEGG pathways with the largest number of proteins. The abscissa represents the name of pathways, and the ordinate represents the number of proteins within each pathway.

As shown in Figure 2B, the top 20 KEGG pathways with the largest number of proteins involved were obtained by KEGG classification annotations. Among them, the top 10 pathways were biosynthesis of antibiotics, biosynthesis of amino acids, carbon metabolism, ABC transporter, phosphotransferase system (PTS), ribosome, starch and sucrose metabolism, quorum sensing, glycolysis/gluconeogenesis, and purine metabolism.

3.3. Analysis of DEPs

Through the differential expression analysis of all the identified proteins between the linalool treatment group and control group, a total of 445 DEPs were screened out, including 211 up-regulated and 234 down-regulated proteins. As shown in Figure 3, the data of the DEPs were converted to volcano plots and heat maps for more intuitionistic comparative analysis.

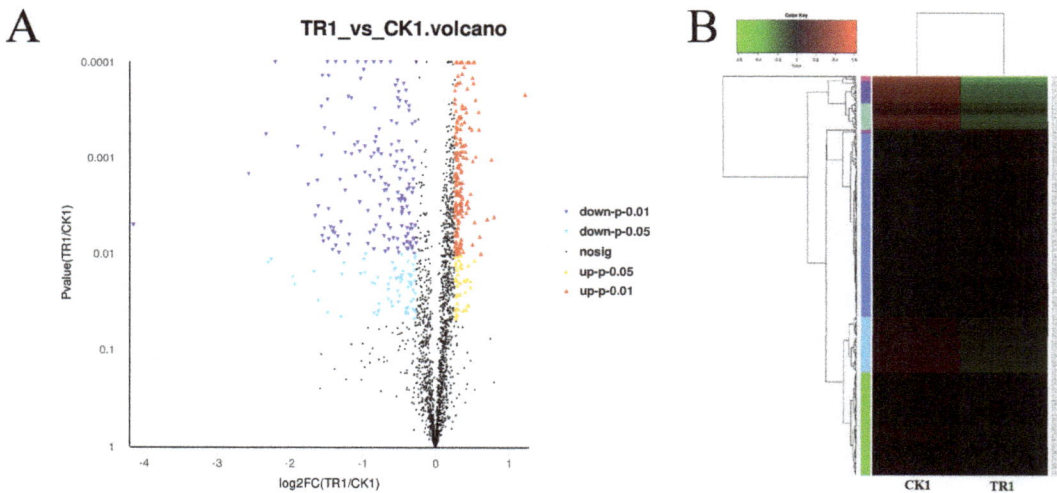

Figure 3. DEPs of LM cells analyzed between the treated group (treated with linalool) and the control group (untreated with linalool). (**A**) Volcano plot of DEPs. The abscissa represents the fold change value of the difference between control and treated samples. The difference value is obtained by dividing the expression level of control sample by treated sample, and this value is logarithmized. The ordinate represents p-value (by the analysis of statistical t-test) of the difference of protein expressions. The smaller the p-value, the more significant the difference in protein expression. Each point represents a specific protein: the yellow point (significantly up-regulated, $p < 0.05$), the red point (significantly up-regulated, $p < 0.01$), the light blue point (significantly down-regulated, $p < 0.05$), the blue point (significantly down-regulated, $p < 0.01$), and the black dots (non-significantly different proteins, $p > 0.05$). (**B**) Heat map of DEPs. The left (CK1, control group) and right columns (TR1, treated group) represent control and treated groups, respectively. Each row represents a protein. Red and green colors represent the high and low expression levels of the protein, respectively. On the left is the dendrogram of protein clustering, and on the right is the name of the protein.

3.4. The GO Enrichment Analysis

The enrichment analysis method is usually used to analyze whether a group of proteins has appeared on a certain functional node in a certain pathway; the aim is to make the annotation analysis from a single protein to a protein set. Enrichment analysis improves the reliability of research and can screen out the biological processes most relevant to biological phenomena. Therefore, GO enrichment analysis was performed to analyze the functional enrichment of differential proteins and clarify the differences between treated and control groups at the functional level.

As shown in Figure 4, a total of 31 significantly enriched functional categories were obtained. Among them, there were 12 categories in "Biological Process", including biological adhesion (GO:0022610), biological regulation (GO:0065007), carbon utilization (GO:0015976), cellular component organization or biogenesis, (GO:0071840), cellular process (GO:0009987), developmental process (GO:0032502), localization (GO:0051179), locomotion (GO:0040011), metabolic process (GO:0008152), multi-organism process (GO:0051704), reproductive process (GO:0022414), and response to stimulus (GO: 0050896). There were 10 categories in "Cell Component", including cell (GO:0005623), cell part (GO:0044464), extracellular region (GO:0005576), extracellular region part (GO:0044421), membrane (GO:0016020), membrane part (GO:0044425), nucleoid (GO:0009295), organelle (GO:0043226), organelle part (GO:0044422), and protein-containing complex (GO:0032991). There were nine categories in "Molecular Function", including antioxidant activity (GO:0016209), binding (GO:0005488), catalytic activity (GO:0003824), molecular carrier activity (GO:0140104), molecular function regulator (GO:0098772), structural molecule activity (GO:0005198), transcription regulator activity (GO:0140110), translation regulator activity (GO:0045182), and transporter activity (GO:0005215).

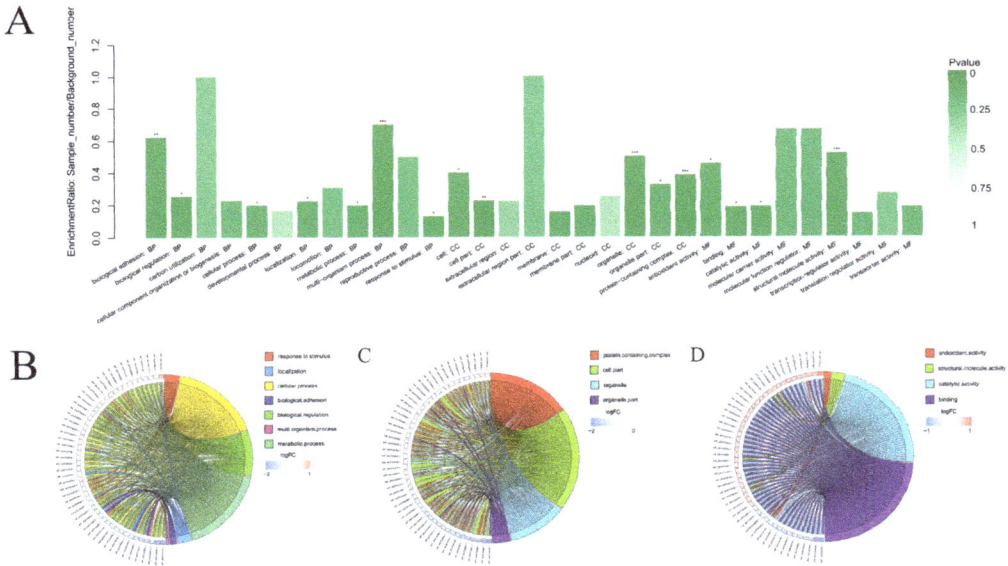

Figure 4. The GO enrichment analyzed between the treated group (treated with linalool) and the control group (untreated with linalool). (**A**) Histogram of GO enrichment of DEPs. Each column represents a GO term, and the abscissa represents the name and category of GO. The ordinate represents the enrichment rate. The color represents the significance of enrichment. $p < 0.05$, $p < 0.01$, $p < 0.001$ are marked as *, **, and ***, respectively. Subfigures (**B**–**D**) are chord diagrams of the GO enrichment of DEPs, respectively showing the (**B**) different proteins participating in specific functions in the three GO categories of biological process, (**C**) cell composition, (**D**) and molecular function.

3.5. The KEGG Enrichment Analysis

In addition to the GO enrichment analysis, KEGG pathway enrichment analysis was also performed to analyze the biological pathways involved in DEPs in this study. As shown in Figure 5, through KEGG enrichment analysis, a total of 60 significantly enriched biological pathways were identified. Among them, there were 6 pathways in the "Cell

Process" category, 6 pathways in the "Environmental Information Processing" category, 3 pathways in the "Human Disease" category, 40 pathways in the "Metabolism" category, and 2 pathways in the "Organic System" category.

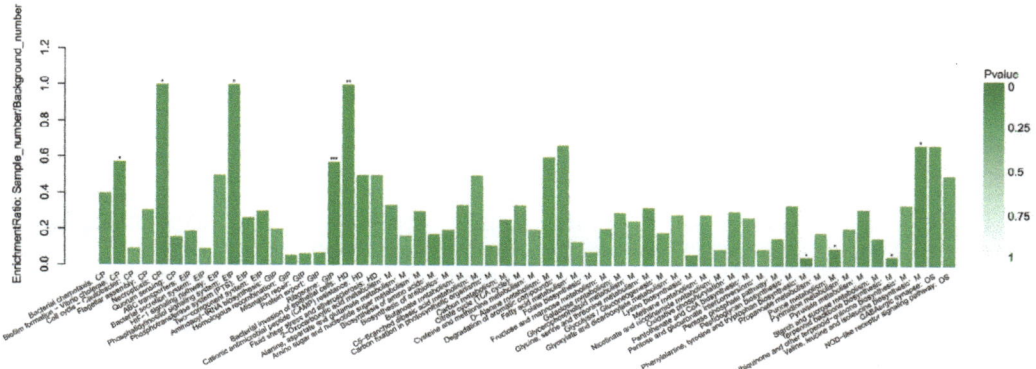

Figure 5. The enrichment of the DEPs in KEGG pathways analyzed between the treated group (treated with linalool) and the control group (untreated with linalool). Each column represents a pathway, and the abscissa represents the name and classification of the pathway: CP (cellular process), EIP (environmental information processing), GIP (genetic information processing), HD (human diseases), M (metabolism), and OS (organismal systems). The height of the column or the ordinate represents the enrichment rate. The color represents the significance of enrichment. $p < 0.05$, $p < 0.01$, and $p < 0.001$ are marked as *, **, and ***, respectively.

3.6. Flow Cytometry Analysis

Five fluorescent dyes (TO, PI, BOX, EB, and CTC) were used to evaluate several vital biological functions in LM cells by flow cytometry analysis. Membrane integrity was evaluated by double staining of TO and PI as shown in Figure 6A. In the control group, 90.4% of the cells were located in plot Q1 (TO+ and PI−), which represented cells with an intact cell membrane. By contrast, in the treated groups, 98.8% of the cells were located in plot Q4 (TO− and PI−), which represented cells with damaged DNA or RNA.

Membrane potential was evaluated by double staining of BOX and PI as shown in Figure 6B. In the control group, 70.8% of the cells were located in plot Q4 (BOX− and PI−), which represented cells with a polarized membrane. By contrast, in the treated groups, 76.1% and 17.6% of the cells were located in plot Q1 (BOX+ and PI−) and plot Q1 (BOX+ and PI+), which represented cells with depolarized nonpermeabilized and permeabilized membranes, respectively. In total, after the treatment of linalool, 93.7% cells became depolarized.

The efflux activity was evaluated by EB staining as shown in Figure 6C. EB− represents the efflux pump functioning properly, while EB+ represents the malfunction of the efflux pump. The percentages of EB+ cells in the control and treated groups were 2% and 85.5%, respectively, and the percentages of EB− cells in these two groups were 96% and 6.97% respectively. These data show that the efflux pump of 85.5% of the cells became malfunctioning after the treatment of linalool.

Respiratory activity was evaluated by CTC staining as shown in Figure 6D. CTC+ and CTC− represent respiratory active and inactive cells, respectively. The percentage of CTC+ cells in the control and treated groups were 99% and 8.53%, and the percentage of CTC− cells in these two groups were 0.96% and 84.6%.

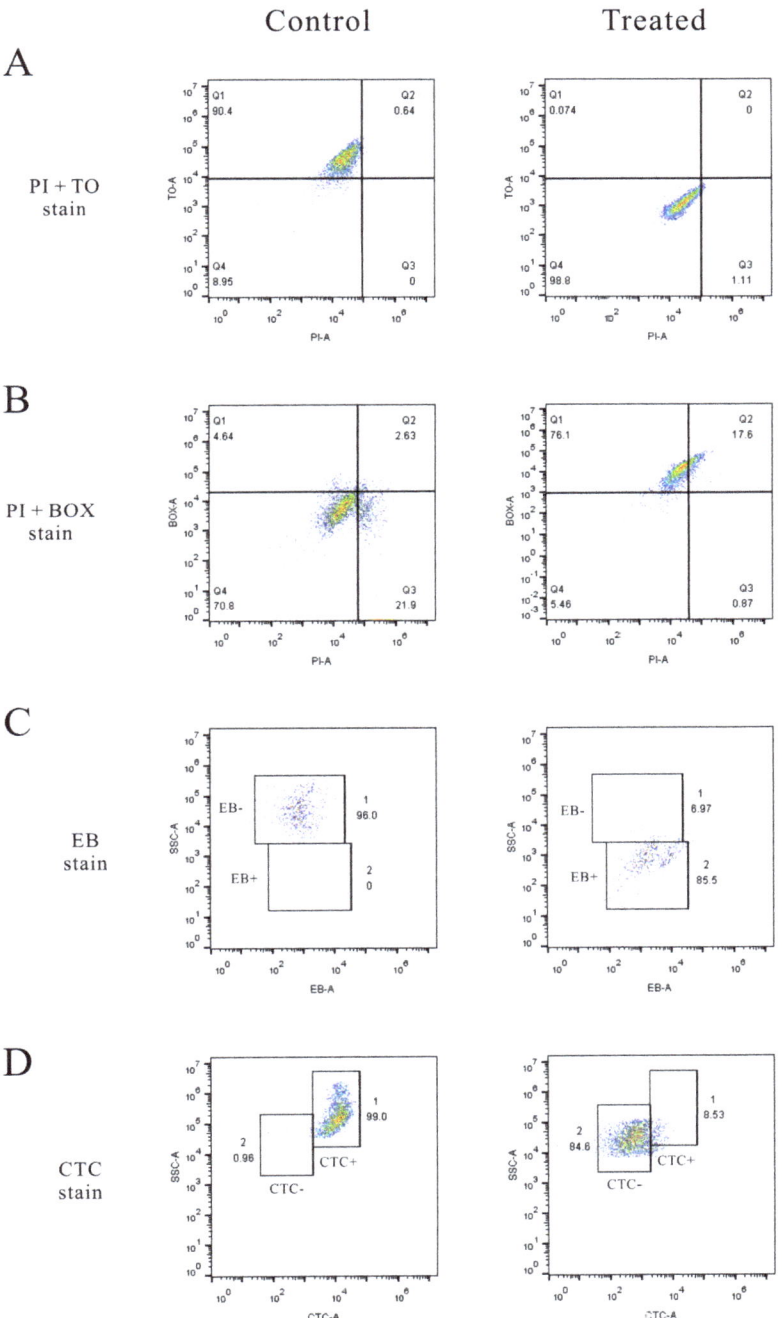

Figure 6. Fluorescence density plots of LM treated (treated) and untreated (control) with linalool stained with (**A**) PI and TO, (**B**) PI and BOX, (**C**) EB, and (**D**) CTC. For subfigures (**A**) and (**B**), the vertical and horizontal axis indicate the fluorescence intensity. The percentages of the cell population in each gate are demonstrated in the four corners of each plot.

4. Discussion

4.1. Proteomic Technology Used in the Antimicrobial Research Field

In this study, iTRAQ-based quantitative proteomics analysis was performed to identify proteins differentially expressed between the linalool treated group and the untreated group. Finally, a total of 445 DEPs were screened out, including 211 up-regulated and 234 down-regulated proteins which participated in different biological functions and pathways.

Except for our study, there have been few other studies focused on the theme of "changes of microbial proteomics after essential oils treatment"; some of these studies are summarized as follows. Xu et al. [32] studied the proteomic changes of *Botrytis cinerea* after tea tree oil treatment, finding a total of 718 DEPs, of which 17 were up-regulated and 701 were down-regulated. These proteins were annotated to 30 GO categories and 133 KEGG pathways, including glycolysis, tricarboxylic acid cycle, and purine metabolism pathways. Hu et al. [33] screened out a total of 745 DEPs of LM after thyme essential oil treatment, of which 246, 45, and 309 proteins were involved in biological processes, cellular components, and molecular functions, respectively. Meanwhile, these proteins participated in 86 KEGG pathways, such as flagella assembly and chemotaxis. Yang et al. [13] demonstrated the proteomic changes of *Klebsiella pneumoniae* after treatment by lavender essential oil. A total of 135 DEPs were found, of which 57 were up-regulated and 78 were down-regulated; they were annotated to 30 GO categories and 133 KEGG pathways. Moreover, 35.78%, 34.01%, 8.44%, and 8.84% of these proteins were involved in cellular processes, metabolic processes, cellular component structure, and stress response, respectively; 58.27%, 17.53%, 13.33%, and 6.17% were involved in the cytoplasm, cell membrane, protein complex, and ribosomal proteins, respectively; and 44.40% and 42.28% were involved in binding and catalytic activity.

Although there are many differences between the above studies and ours, there are also some similarities, such as changes in glycolysis and other metabolic pathways. These differences might be caused by different proteomic technologies, types of essential oils, and the different microbial strains used. Proteomic technology has been proved to be an efficient, fast, and useful approach to identify differentially expressed proteins in various types of microorganisms.

4.2. Further Analysis of Important Functions in GO Enrichment Analysis

Several categories relevant to "cell component" were enriched as shown in Figure 4A. Firstly, the enrichment of "cell (GO:0005623)" and "cell part (GO:0044464)" indicated that the structure and component of bacterial cells were altered significantly after linalool treatment. Secondly, the enrichment of "membrane (GO:0016020)" and "membrane part (GO:0044425)" was in accordance with the results of SEM and TEM in our previous study, which showed obvious damage of the cell membrane [15]. Thirdly, the enrichment of "nucleoid (GO:0009295), organelle (GO:0043226), and organelle part (GO:0044422)" echoed the data of the flow cytometry analysis, which proved that the DNA or RNA (main part of the nucleoid) might be destroyed by linalool treatment; meanwhile, "organelle part" also echoed the enrichment of the "ribosome pathway (ko03010)" in KEGG analysis. To sum up, all the above data implied that the cell membrane, nucleoid, and ribosome may be the potential targets of linalool.

As shown in Figure 4A, some categories related to "molecular function" were also enriched. Initially, the enrichment of "structure molecule activity (GO:0005198)" meant that the action of molecules which contributes to the structural integrity or assembly of a complex significantly changed. Furthermore, the enrichment of both "transcription regulator activity (GO:0140110)" and "translation regulator activity (GO:0045182)" indicated the gene expression and polypeptide synthesis were significantly influenced by linalool, which was further evidence showing that the nucleoid and ribosome might be the potential targets.

Some categories relevant to "biological process" were also enriched as shown in Figure 4A. Primarily, the enrichment of "response to stimulus (GO:0050896)" indicated that cells may produce many adaptive responses to meet the challenge of the stimulus (linalool). Further, the enrichment of "biological adhesion (GO:0022610)" might be related to two proteins (InlA and InlB) as shown in the "bacterial invasion of epithelial cells (ko05100)" pathway in KEGG analysis. These two proteins are surface invasions, which could promote the uptake of LM by host cells [34] and enhance the adherence of LM to a glass surface [35]. Moreover, the enrichment of "reproductive process (GO:0022414)" suggested that linalool treatment might influence the reproduction of cells, which was also shown in our previous study [15].

4.3. Further Analysis of Important Pathways in KEGG Enrichment Analysis

First of all, as shown in the "peptidoglycan biosynthesis (ko00550)" pathway in Figure S1, several proteins related to peptidoglycan biosynthesis were down-regulated, which indicated that the biosynthesis of peptidoglycan was reduced by linalool treatment. Two kinds of penicillin-binding proteins (PBPs), PBP 2 and PBP 1a, were down-regulated. PBPs are a major class of enzymes related to peptidoglycan synthesis, which were identified as targets of β-lactam antibiotics (such as penicillin); they inactivate the crosslinking domains of peptidoglycan synthesis covalently [36]. Peptidoglycan is a vital component of cell walls, maintaining the morphology and viability of bacterial cells [37]. It has been proven that peptidoglycan is the target of many antimicrobial drugs, such as penicillin [36] and cephalosporin [38]. Thus, our results imply that peptidoglycan and cell walls might be another important drug target of linalool.

Next, two peptidoglycan enzymes, AmiA and AmiC, were down-regulated as shown in Figure S2. AmiA and AmiC are N-acetylmuramyl-l-alanine amidases that move side chains away from bacterial peptidoglycan through cleaving the amide bond between peptides and N-acetylmuramic acids [39], which are crucial for septal splitting and the separation of daughter cells [40]. Thus, the down-regulation of these two proteins indicated that the cell division was inhibited by linalool treatment, and this phenomenon might also prove the changing of the "reproductive process" in the GO enrichment analysis.

Moreover, many proteins relevant to the structure and function of ribosome were significantly up- or down-regulated as shown in the "ribosome (ko03010)" pathway in Figure S3, which indicates that the function of ribosome was obviously affected by linalool treatment. These were consistent with our results in GO enrichment analysis, which again suggested ribosomes were the target of linalool.

To sum up, based on our results from the GO and KEGG enrichment analysis, cell membranes, cell walls, nucleoids, and ribosomes might be the main targets of linalool against LM. These results also proved the multi-target effects of linalool, which is one of its most important advantages compared to traditional antibiotics.

5. Conclusions

In this study, iTRAQ-based quantitative proteomics sequencing was carried out to unravel the mode of action of linalool against LM at the protein level. A total of 445 DEPs (including 211 up-regulated and 234 down-regulated proteins) were screened out between the linalool treatment group and the control group. GO and KEGG enrichment analysis implied that cell membranes, cell walls, nucleoids, and ribosomes might be the targets of linalool against LM, which was also supported by the results of the flow cytometry analysis. In the future, we will focus on each of these targets and investigate the specific mechanisms. Meanwhile, the application of linalool for the prevention of LM in foods or food facilities should also be explored.

Supplementary Materials: The following are available online at https://www.mdpi.com/article/10.3390/foods10102449/s1, Figure S1: Peptidoglycan biosynthesis pathway from the KEGG analysis. Figure S2: AmiA and AmiC annotation from the KEGG analysis. Figure S3: Ribosome pathway from the KEGG annotation analysis.

Author Contributions: Methodology, Z.G.; validation, J.G.; investigation, W.Z. and T.Z.; writing—original draft preparation, J.G.; writing—review and editing, Z.G.; visualization, T.L.; supervision,

Z.G. and J.G.; project administration, Z.G.; funding acquisition, Z.G. All authors have read and agreed to the published version of the manuscript.

Funding: This research was funded by the National Natural Science Foundation of China (32073020), the Key Projects of Hunan Education Department (20A238), the Changsha Municipal Natural Science Foundation (kq2014070), the Key Laboratory of Agro-Products Processing, and the Ministry of Agriculture and Rural Affairs of P. R. China (S2021KFKT-22).

Conflicts of Interest: The authors declare no conflict of interest.

References

1. King, T.; Cole, M.; Farber, J.M.; Eisenbrand, G.; Zabaras, D.; Fox, E.M.; Hill, J.P. Food safety for food security: Relationship between global megatrends and developments in food safety. *Trends Food Sci. Technol.* **2017**, *68*, 160–175. [CrossRef]
2. Radoshevich, L.; Cossart, P. *Listeria monocytogenes*: Towards a complete picture of its physiology and pathogenesis. *Nat. Rev. Microbiol.* **2018**, *16*, 32–46. [CrossRef]
3. Freitag, N.E.; Port, G.C.; Miner, M.D. *Listeria monocytogenes*—From saprophyte to intracellular pathogen. *Nat. Rev. Microbiol.* **2009**, *7*, 623–628. [CrossRef]
4. Pizarro-Cerda, J.; Cossart, P. *Listeria monocytogenes*: Cell biology of invasion and intracellular growth. *Microbiol. Spectr.* **2018**, *6*, 6.6.05. [CrossRef]
5. European Food Safety Authority, European Centre for Disease Prevention and Control. The European Union One Health 2019 Zoonoses Report. *EFSA J.* **2021**, *19*, e06406.
6. Maury, M.M.; Bracq-Dieye, H.; Huang, L.; Vales, G.; Lavina, M.; Thouvenot, P.; Disson, O.; Leclercq, A.; Brisse, S.; Lecuit, M. Hypervirulent *Listeria monocytogenes* clones' adaption to mammalian gut accounts for their association with dairy products. *Nat. Commun.* **2019**, *10*, 1–13. [CrossRef]
7. Fan, Z.; Xie, J.; Li, Y.; Wang, H. Listeriosis in mainland China: A systematic review. *Int. J. Infect. Dis.* **2019**, *81*, 17–24. [CrossRef]
8. Smith, A.M.; Tau, N.P.; Smouse, S.L.; Allam, M.; Ismail, A.; Ramalwa, N.R.; Disenyeng, B.; Ngomane, M.; Thomas, J. Outbreak of *Listeria monocytogenes* in South Africa, 2017–2018: Laboratory activities and experiences associated with whole-genome sequencing analysis of isolates. *Foodborne Pathog. Dis.* **2019**, *16*, 524–530. [CrossRef]
9. Charlier, C.; Disson, O.; Lecuit, M. Maternal-neonatal listeriosis. *Virulence* **2020**, *11*, 391–397. [CrossRef]
10. Guo, J.; Gao, Z.; Li, G.; Fu, F.; Liang, Z.; Zhu, H.; Shan, Y. Antimicrobial and antibiofilm efficacy and mechanism of essential oil from *Citrus Changshan-huyou YB chang* against *Listeria monocytogenes*. *Food Control* **2019**, *105*, 256–264. [CrossRef]
11. Awad, A.H.; Parmar, A.; Ali, M.R.; El-Mogy, M.M.; Abdelgawad, K.F. Extending the shelf-life of fresh-cut green bean pods by ethanol, ascorbic acid, and essential oils. *Foods* **2021**, *10*, 1103. [CrossRef]
12. Guo, J.; Hu, X.; Gao, Z.; Li, G.; Fu, F.; Shang, X.; Liang, Z.; Shan, Y. Global transcriptomic response of *Listeria monocytogenes* exposed to Fingered Citron (*Citrus medica* L. var. *sarcodactylis* Swingle) essential oil. *Food Res. Int.* **2021**, *143*, 110274. [CrossRef]
13. Yang, S.-K.; Yusoff, K.; Thomas, W.; Akseer, R.; Alhosani, M.S.; Abushelaibi, A.; Lai, K.-S. Lavender essential oil induces oxidative stress which modifies the bacterial membrane permeability of carbapenemase producing *Klebsiella pneumoniae*. *Sci. Rep.* **2020**, *10*, 1–14. [CrossRef]
14. US Food and Drug Administration. *Drug Administration Code of Federal Regulations Title 21*; 21CFR20157; Department of Health and Human Services, Ed.; US Food and Drug Administration: Washington, DC, USA, 2014.
15. Gao, Z.; Van Nostrand, J.D.; Zhou, J.; Zhong, W.; Chen, K.; Guo, J. Anti-*Listeria* activities of linalool and its mechanism revealed by comparative transcriptome analysis. *Front. Microbiol.* **2019**, *10*, 2947. [CrossRef]
16. Guo, J.-J.; Gao, Z.-P.; Xia, J.-L.; Ritenour, M.A.; Li, G.-Y.; Shan, Y. Comparative analysis of chemical composition, antimicrobial and antioxidant activity of citrus essential oils from the main cultivated varieties in China. *Lwt* **2018**, *97*, 825–839. [CrossRef]
17. Gao, Z.; Zhong, W.; Chen, K.; Tang, P.; Guo, J. Chemical composition and anti-biofilm activity of essential oil from *Citrus medica* L. var. *sarcodactylis* Swingle against *Listeria monocytogenes*. *Ind. Crop. Prod.* **2020**, *144*, 112036. [CrossRef]
18. Xu, Y.-J. Foodomics: A novel approach for food microbiology. *TrAC Trends Anal. Chem.* **2017**, *96*, 14–21. [CrossRef]
19. Misra, B.B.; Langefeld, C.; Olivier, M.; Cox, L.A. Integrated omics: Tools, advances and future approaches. *J. Mol. Endocrinol.* **2019**, *62*, R21–R45. [CrossRef]
20. Walsh, A.M.; Crispie, F.; Claesson, M.J.; Cotter, P.D. Translating omics to food microbiology. *Ann. Rev. Food Sci. Technol.* **2017**, *8*, 113–134. [CrossRef]
21. Gajdošik, M.Š.; Andjelković, U.; Gašo-Sokač, D.; Pavlović, H.; Shevchuk, O.; Martinović, T.; Clifton, J.; Josić, D. Proteomic analysis of food borne pathogens following the mode of action of the disinfectants based on pyridoxal oxime derivatives. *Food Res. Int.* **2017**, *99*, 560–570. [CrossRef]
22. Zhao, Y.-L.; Zhou, Y.-H.; Chen, J.-Q.; Huang, Q.-Y.; Han, Q.; Liu, B.; Cheng, G.-D.; Li, Y.-H. Quantitative proteomic analysis of sub-MIC erythromycin inhibiting biofilm formation of *S. suis* in vitro. *J. Proteom.* **2015**, *116*, 1–14. [CrossRef]
23. Hesketh, A.; Deery, M.J.; Hong, H.-J. High-resolution mass spectrometry based proteomic analysis of the response to vancomycin-induced cell wall stress in *Streptomyces coelicolor* A3(2). *J. Proteome Res.* **2015**, *14*, 2915–2928. [CrossRef]

24. Ma, W.; Zhang, D.; Li, G.; Liu, J.; He, G.; Zhang, P.; Yang, L.; Zhu, H.; Xu, N.; Liang, S. Antibacterial mechanism of daptomycin antibiotic against *Staphylococcus aureus* based on a quantitative bacterial proteome analysis. *J. Proteom.* **2017**, *150*, 242–251. [CrossRef]
25. Yuan, P.; He, L.; Chen, D.; Sun, Y.; Ge, Z.; Shen, D.; Lu, Y. Proteomic characterization of *Mycobacterium tuberculosis* reveals potential targets of bostrycin. *J. Proteom.* **2020**, *212*, 103576. [CrossRef]
26. Li, L.; Tian, Y.; Yu, J.; Song, X.; Jia, R.; Cui, Q.; Tong, W.; Zou, Y.; Li, L.; Yin, L. iTRAQ-based quantitative proteomic analysis reveals multiple effects of Emodin to *Haemophilus parasuis*. *J. Proteom.* **2017**, *166*, 39–47. [CrossRef]
27. Sun, L.; Chen, H.; Lin, W.; Lin, X. Quantitative proteomic analysis of *Edwardsiella tarda* in response to oxytetracycline stress in biofilm. *J. Proteom.* **2017**, *150*, 141–148. [CrossRef]
28. Zieske, L.R. A perspective on the use of iTRAQ™ reagent technology for protein complex and profiling studies. *J. Exp. Bot.* **2006**, *57*, 1501–1508. [CrossRef] [PubMed]
29. Mahoney, D.W.; Therneau, T.M.; Heppelmann, C.J.; Higgins, L.; Benson, L.M.; Zenka, R.M.; Jagtap, P.; Nelsestuen, G.L.; Bergen, H.R., III; Oberg, A.L. Relative quantification: Characterization of bias, variability and fold changes in mass spectrometry data from iTRAQ-labeled peptides. *J. Proteome Res.* **2011**, *10*, 4325–4333. [CrossRef]
30. De Sousa Guedes, J.P.; de Souza, E.L. Investigation of damage to *Escherichia coli*, *Listeria monocytogenes* and *Salmonella Enteritidis* exposed to *Mentha arvensis* L. and *M. piperita* L. essential oils in pineapple and mango juice by flow cytometry. *Food Microbiol.* **2018**, *76*, 564–571. [CrossRef]
31. Zhang, C.; Chen, X.; Xia, X.; Li, B.; Hung, Y.-C. Viability assay of *E. coli* O157: H7 treated with electrolyzed oxidizing water using flow cytometry. *Food Control* **2018**, *88*, 47–53. [CrossRef]
32. Xu, J.; Shao, X.; Wei, Y.; Xu, F.; Wang, H. iTRAQ proteomic analysis reveals that metabolic pathways involving energy metabolism are affected by tea tree oil in *Botrytis cinerea*. *Front. Microbiol.* **2017**, *8*, 1989. [CrossRef]
33. Hu, W.; Feng, K.; Xiu, Z.; Jiang, A.; Lao, Y. Tandem mass tag-based quantitative proteomic analysis reveal the inhibition mechanism of thyme essential oil against flagellum of *Listeria monocytogenes*. *Food Res. Int.* **2019**, *125*, 108508.
34. Phelps, C.C.; Vadia, S.; Arnett, E.; Tan, Y.; Zhang, X.; Pathak-Sharma, S.; Gavrilin, M.A.; Seveau, S. Relative roles of listeriolysin O, InlA, and InlB in *Listeria monocytogenes* uptake by host cells. *Infect. Immun.* **2018**, *86*, e00555-18. [CrossRef] [PubMed]
35. Chen, B.-Y.; Kim, T.-J.; Silva, J.L.; Jung, Y.-S. Positive correlation between the expression of inlA and inlB genes of *Listeria monocytogenes* and its attachment strength on glass surface. *Food Biophys.* **2009**, *4*, 304–311. [CrossRef]
36. Taguchi, A.; Welsh, M.A.; Marmont, L.S.; Lee, W.; Sjodt, M.; Kruse, A.C.; Kahne, D.; Bernhardt, T.G.; Walker, S. FtsW is a peptidoglycan polymerase that is functional only in complex with its cognate penicillin-binding protein. *Nat. Microbiol.* **2019**, *4*, 587–594. [CrossRef] [PubMed]
37. Gautam, A.; Vyas, R.; Tewari, R. Peptidoglycan biosynthesis machinery: A rich source of drug targets. *Crit. Rev. Biotechnol.* **2011**, *31*, 295–336. [CrossRef]
38. Strehl, E.; Kees, F. Pharmacological Properties of Parenteral Cephalosporins. *Drugs* **2000**, *59*, 9–18. [CrossRef]
39. Heidrich, C.; Templin, M.F.; Ursinus, A.; Merdanovic, M.; Berger, J.; Schwarz, H.; De Pedro, M.A.; Höltje, J.V. Involvement of N-acetylmuramyl-L-alanine amidases in cell separation and antibiotic-induced autolysis of *Escherichia coli*. *Mol. Microbiol.* **2001**, *41*, 167–178. [CrossRef]
40. Stohl, E.A.; Lenz, J.D.; Dillard, J.P.; Seifert, H.S. The gonococcal NlpD protein facilitates cell separation by activating peptidoglycan cleavage by AmiC. *J. Bacteriol.* **2016**, *198*, 615–622. [CrossRef] [PubMed]

Article

High-Pressure Carbon Dioxide Use to Control Dried Apricot Pests, *Tribolium castaneum* and *Rhyzopertha dominica*, and Assessing the Qualitative Traits of Dried Pieces of Treated Apricot

Reza Sadeghi [1,*], Fereshteh Heidari [1], Asgar Ebadollahi [2,*], Fatemeh Azarikia [3], Arsalan Jamshidnia [1] and Franco Palla [4,*]

1. Department of Entomology and Plant Pathology, College of Aburaihan, University of Tehran, Tehran 3391653755, Iran; f.heidari4234@gmail.com (F.H.); jamshidnia@ut.ac.ir (A.J.)
2. Department of Plant Sciences, Moghan College of Agriculture and Natural Resources, University of Mohaghegh Ardabili, Ardabil 5619936514, Iran
3. Department of Food Technology, College of Aburaihan, University of Tehran, Tehran 3391653755, Iran; azarikia@ut.ac.ir
4. Department of Biological, Chemical and Pharmaceutical Sciences and Technologies, University of Palermo, 38-90123 Palermo, Italy
* Correspondence: rsadeghi@ut.ac.ir (R.S.); ebadollahi@uma.ac.ir (A.E.); franco.palla@unipa.it (F.P.)

Citation: Sadeghi, R.; Heidari, F.; Ebadollahi, A.; Azarikia, F.; Jamshidnia, A.; Palla, F. High-Pressure Carbon Dioxide Use to Control Dried Apricot Pests, *Tribolium castaneum* and *Rhyzopertha dominica*, and Assessing the Qualitative Traits of Dried Pieces of Treated Apricot. *Foods* 2021, 10, 1190. https://doi.org/10.3390/foods10061190

Academic Editors: Lisa Pilkington, Siew-Young Quek and Susana Casal

Received: 10 April 2021
Accepted: 23 May 2021
Published: 25 May 2021

Publisher's Note: MDPI stays neutral with regard to jurisdictional claims in published maps and institutional affiliations.

Copyright: © 2021 by the authors. Licensee MDPI, Basel, Switzerland. This article is an open access article distributed under the terms and conditions of the Creative Commons Attribution (CC BY) license (https://creativecommons.org/licenses/by/4.0/).

Abstract: One of the new ways of warehouse pest control is the carbon dioxide treatment, which had no residues on the target products. In the present research, at first, CO_2 gas was applied to control two important pest species infesting dried apricots. Dry apricots infested with adults of *Tribolium castaneum* (Herbst) or *Rhyzopertha dominica* (F.) were exposed to CO_2 gas pressures correspond to 9.1, 16.7, 23.1, 28.6, and 33.4 mol% for 24 h. The results showed higher mortality rates with increasing the gas pressures in all the experiments. The minimum and maximum losses of the pests were determined at concentrations of 9.1 and 33.4 mol%, respectively. Evaluation of CO_2 gas effects on the quality characteristics of dried apricots showed no impacts on the color, brittleness, hardness, sweetness, sourness, and general acceptance of products. CO_2 gas treatments at the concentration of 33.4 mol% showed no significant influences on the chemical features of dried apricots, including pH, acidity, Brix, humidity percentage, reducing sugar, and total sugar. It was concluded that CO_2 gas had the potential to control *T. castaneum* and *R. dominica* in warehouses of dried apricots, without any significant impacts on product qualities.

Keywords: apricot; CO_2 gas; qualitative traits; warehouse pest

1. Introduction

Apricot, *Prunus armeniaca* L. (Rosales: Rosaceae), is considered as one of the most delicious fruits in regions with temperate climates. It contains saccharides, organic acids, minerals, vitamins, and polyphenols and has antioxidant properties [1]. Apricot is served fresh or as dried or frozen fruit, compote, extract, and jam [2]. According to the Food and Agriculture Organization (FAO) statistics, Turkey, Uzbekistan, and Iran ranked 1st to 3rd in the world in 2018 for producing 750,000, 493,842, and 342,479 tons of apricot, respectively [3]. Dried fruits are stored in warehouses under specific controlled conditions to maintain their quality and marketability, besides preventing pest infestation. Millions of tons of crops and dried fruits are annually lost because of the damage created by storage pests and the non-observance of scientific storage principles. Heavy quantitative, qualitative, and hygienic damages may be inflicted on storage products by insects [4]. To be supplied to the world markets, agricultural products must satisfy the necessary standards such as acceptable taste and color and particularly moisture content without any infestations of insects [5].

Conventional pest management practices provide such advantages as ease of operations and low costs; yet, overuse of synthetic chemicals involves some disadvantages such as prolonging slow operations, producing environmentally polluting and ozone-depleting wastes, imposing health risks on the operators, having adverse impacts on product quality, and so on [6–8]. For instance, methyl bromide has been successfully applied to control a variety of insects over the years, but due to the hazardous environmental impacts its use has been limited by international subsidiary organizations until 2015 [9]. Thus, it is necessary to replace synthetic insecticides with environmentally friendly agents concerning the problems related to chemical insecticides, besides the economic importance of pest control [10]. In previous publications, as an alternative to synthetic pesticides, interests have been focalized on plant essential oils [11,12].

Being a non-flammable and odorless gas with a body mass of 1.5 times the air mass, carbon dioxide (CO_2) had no residues on target products while also being toxic to insects [13]. A wide range of pests, along with their different growth stages, can be controlled by CO_2 gas during the storage of dried fruits [14]. An increase in the amount of CO_2 would lead to enhancement of the respiratory rates of insects. Therefore, in warehouses, insects can be killed even at low CO_2 concentrations, which results in the disruption of their breathing regulations when it enters their body tissues and fluid organs [15]. Additionally, respiratory pores of insects may be kept open at the concentration of 35% CO_2, which can force extra water uptakes of their cellular tissues and impair the metabolism of organs by reducing amounts of triglycerides; these conditions would consequently lead to their mortalities [16]. Effectiveness of CO_2 in the control of stored-product insect pests was assessed in the previous studies. For example, Cheng et al. [17] used a mixture of 2% oxygen and 18% carbon dioxide, which could stop all the growth stages of cowpea weevil, *Callosobruchus maculatus* F. In their study, Valizadegan et al. [18] were able to control four species of storage pests with a combination of phosphine and carbon dioxide, which provided the advantage of reduced phosphine consumption. Complete controls of the insects were achieved within 7, 10, and 11 weeks after 24 h-treatment. Carbon monoxide and dioxide efficacies were investigated by Dhouibi et al. [19] for controlling flour and carob moths (*Ephestia kuehniella* Zeller and *Ectomyelois ceratoniae* Zeller, respectively) infesting organic dates. Mortality of all tested insects was achieved by 5.5 and 2.8 kg/ m^3 of CO_2 fumigations after 20 and 48 h under laboratory and field conditions, respectively.

The color lab space was designed in a way to be very similar to what is perceived by human vision [20]. In the image processing technique, the parameters of L*, a*, and b* indicate brightness, redness, and yellowness of products, respectively, which can be changed in the treatments by pest control and preserving agents [21]. For example, Sadeghi et al. [22] reported increasing L*, a*, and b* variations for figs and raisins with enhancing microwave powers and timings in the sawtoothed grain beetle (*Oryzaephilus surinamensis* L.) management. Additionally, Inserra et al. [23] observed an increase in L* and b* and a reduction in a* parameters of apricots treated with sulfur before drying.

Although the red flour beetle (*Tribolium castaneum* Herbst) and the lesser grain borer (*Rhyzopertha dominica* (F.)) are among the cosmopolitan insect pests of stored flour and grains, they can damage dried material of animal and plant origin such as dried fruits [24,25]. This study aimed to (1) determine the effect of high-pressure CO_2 on the mortality of two cosmopolitan stored-products insect pests *T. castaneum* and *R. dominica* and (2) investigate possible changes in sensory (organoleptic) and chemical properties of dried apricot affected after treatment by CO_2.

2. Materials and Methods

2.1. Dried Fruit and Rearing Insect Pests

The stored product used in present investigation was dried apricot (*Prunus armeniaca* L., Shahroudi cultivar) purchased from the wholesale dried fruit market in Tehran City, Iran. *R. dominica* and *T. castaneum*, which had been previously reared for ten generations, were prepared in the Toxicology Laboratory of the Department of Entomology and Plant

Pathology, College of Aburaihan, University of Tehran, Iran. The 1.5 kg of the two food compositions of broken cereal [26] and intact wheat grains [27] were utilized for breeding *T. castaneum* and *R. dominica*, respectively. Then, 200 insects of each pest (without sex separation) were released in 3-L plastic containers to be reared at the temperature of 27 ± 2 °C, relative humidity of 65% ± 5%, and light: dark of 10:14 h [28]. Afterward, a brush was used to remove 50 insects from the permanent foodstuff and place them on the rearing food in new containers. Adults were then separated and transferred to the main dish to avoid generational intermingling after a week of mating and spawning. After 27 days, the next adults of the insects appeared and 1–10 day-old adults were used in the experiments.

2.2. Fumigation of Carbon Dioxide

The pressures of CO_2 gas, which was provided by 20-L capsules, were determined by the equipped pressure gauge during the preliminary tests. The experiment was laid out in a completely randomized design (CRD) with ten replications. Disposable paper cups, with the diameter and height of 7.5 and 8.7 cm, respectively, were used to hold 30 g of the dried apricot samples contaminated with twenty pests. After transferring the containers of each treatment to a cylindrical autoclave (Kavush Mega Medical Instrument Co., Tehran, Iran) with a 10-L stainless steel tank equipped with a pressure gauge, CO_2 inlet and outlet valves, and closable steel lids to confine CO_2, the lids were covered with lace and elastic bands. Air inside the autoclave was allowed to completely exit by opening the outlet valve after 20 s of fumigation. Then, the valve closed and fumigation was prolonged at the specified pressures from 0.1 to 0.5 bar correspond to 9.1, 16.7, 23.1, 28.6, and 33.4 mol%. Finally, 24 h after CO_2 fumigation and aeration, the mortality of adult insects was counted. The control treatments underwent no pressure of CO_2 gas.

2.3. Qualitative Properties

The samples of the dried apricot were studied for organoleptic properties, including color, aroma, sweetness, sourness, brittleness, hardness, and general acceptance, before and after CO_2 treatments. Ten expert evaluators (5 males and 5 females) confirmed in the taste validity test according to the standards [29] were employed to assess the characteristics. The sensory properties of the apricot specimens were judged by the evaluators using the 7-point hedonic test. The numerical scales of 1–7 (higher qualitative levels with increasing numbers) were recorded in their given forms expressing the intensity of each qualitative attribute. The treated samples were then compared with the five control samples poured into the plastic containers. The evaluators were also provided with some disposable cups of mineral water to avoid interfering with the tastes of the samples between each detection step. These experiments were also arranged in a completely randomized design (CRD) in ten replications.

A digital camera (OLIMPUS VR-310, Tokyo, Japan) was employed to take photos of the pieces of apricots. An artificial lighting system was utilized to provide a uniform lighting condition for the specimens. It was done by using a white fluorescent lamp mounted under the ceiling of our special chamber, the walls of which were covered with a black cover. Photographing was performed at a distance of 30 cm from the specimens put against a white background [22]. The images were captured at 959 pixels × 1280 pixels and a resolution of 96 dpi in Jpeg format and RGB color space. Image 1.50 v software was applied to process the images captured by the digital camera.

2.4. Determination of Chemical Properties

First, the 3-g samples of the apricot pieces, in five replications and pretreated with CO_2 gas at the concentration of 33.4 mol%, were stirred in 20 mL of distilled water using a magnetic stirrer at 80 °C. After 2 h, the liquid part was separated by filter paper [23]. To determine pH and Brix number, a pH meter and a refractometer were employed, respectively.

To specify the acidity, 10 mL of the prepared solution in five replications was titrated with NaOH (0.1 M) until obtaining a pale pink color (pH = 8.3). The results were reported in terms of dominant acid percentage: malic acid [23].

To delineate the moisture content, 3 g of the samples (in three replications) were placed in a glass plate heated by an oven to 80 °C for 18 h, and the percentage of moisture content was determined by dividing the difference of the sample weights before and after heating [30].

To determine the reducing sugar, 3 g of the sample (in three replications) were first mixed with 40 mL of distilled water at 80 °C and was filtered through filter paper after 2 h. Then, 5 mL of Fehling A and Fehling B together with a magnet and about 10 mL of distilled water were poured into a 100-mL Erlenmeyer flask. Afterward, the Erlen was placed on a magnetic stirrer with a heater and slowly titrated with the sugar solution until fading the blue color and appearing the brick color [31].

To specify total sugar, 25 mL of the filtered solution was first poured into a 100-mL volumetric flask and 5 mL of concentrated hydrochloric acid (37%) was added to it to be then heated in a boiling water bath (Bain Marie) at 65 °C for 10 min. After reaching ambient temperature, 2 drops of phenol phethalene reagent were added to the mixture, which was then neutralized with 45% concentrated sodium hydroxide (0.1) to give a pale pink color. Finally, it was volumized with distilled water and then titrated with the prepared sugar solution according to the method described for reducing sugar [31]. This experiment was also performed in three replications. All experiments about the evaluation of chemical properties of dried apricots were arranged in a completely randomized design (CRD).

2.5. Statistical Analysis

The Kolmogorov-Smirnov test was used to check the normality of the data. Analysis of variance (ANOVA) was used to determine the statistical effect of CO_2 pressures on insect pests. Differences between means were assessed using the Tukey HSD test at $p < 0.05$. When any death was observed in the control groups, mortality percentages were corrected using Abbott's formula [32]. Lethal concentration values (LC_{50} and LC_{90} with their 95% fiducial limits) and the regression line particulars were calculated using Probit analysis. All statistical analyses were performed with SPSS version 25 (IBM, Chicago, IL, USA). Evaluation of chemical properties (pH, acidity, Brix, total sugar, reducing sugar, and moisture) between treated and untreated dried apricot samples were compared using the *t*-test. A non-parametric Friedman's test used for sensory assessment.

3. Results

The adults of *R. dominica* and *T. castaneum* were highly susceptible to the fumigation of CO_2. According to the results of the analysis of variance, the mortality of *R. dominica* and *T. castaneum* adults was statistically affected by different concentrations of CO_2. As can be seen in Table 1, the increasing CO_2 concentration were associated with the enhanced mortality of the both insects. The CO_2 concentration of 33.4 mol% caused 90 and 78% mortality of the adults of *R. dominica* and *T. castaneum*, respectively (Table 1).

Along with the high mortality of *R. dominica* at all tested concentrations of CO_2, the calculated LC_{50} value (20.19 mol%) for this pest was also lower than that corresponding value for *T. castaneum* (27.02 mol%). However, based on the overlapping of their fiducial limits, the susceptibility of these insects to CO_2 was not significantly different (Table 2). Additionally, according to the R^2 values in Table 2, there was a direct and positive relationship between the tested CO_2 concentrations and the observed mortality of both insects.

Table 1. Means ± standard errors of the mortality rates of T. castaneum and R. dominica at different CO_2 pressures.

Concentration (mol%)	Mortality Percentage + SE	
	T. castaneum	R. dominica
Control = 0	1.00 ± 0.667 [a]	3.50 ± 1.302 [a]
9.1	5.50 ± 1.167 [a]	23.00 ± 1.106 [b]
16.7	9.50 ± 1.572 [a]	44.00 ± 2.667 [c]
23.1	27.50 ± 3.184 [b]	57.50 ± 2.713 [d]
28.6	57.00 ± 1.700 [c]	80.00 ± 2.108 [e]
33.4	78.00 ± 3.000 [d]	90.50 ± 3.114 [f]
ANOVA:	F = 223.54 df = 5, 54 p = 0 < 0.001 *	F = 210.23 df = 5, 54 p = 0 < 0.001 *

Values with different letters within each column are statistically different according to the Tukey's test at $p < 0.05$. Asterisks indicate significant effects of CO_2 concentrations on mortality of insect pests, according to the results of the analysis of variance (ANOVA). SE is standard error.

Table 2. Probit analysis of the toxicity of CO_2 on the adult-insects of R. dominica and T. castaneum.

Insect	LC_{50} with 95% Fiducial Limits (mol%)	LC_{90} with 95% Fiducial Limits (mol%)	Intercept ± SE	Slope ± SE	χ^2 (df = 3)	Sig.	R^2 Value
R. dominica	20.19 (18.37–22.19)	54.45 (38.41–131.89)	−3.88 ± 0.43	2.97 ± 0.32	5.58	0.13	0.95
T. castaneum	27.02 (18.42–104.51)	48.26 (32.83–5320.19)	−7.21 ± 0.67	5.03 ± 0.49	26.19	0 < 0.001	0.87

Number of tested adults for each species was 500. Sig. is significant.

Table 3 shows the means of the sensory characteristics of the dried apricots and their comparison via the Friedman's test. The results revealed that the different concentrations of CO_2 gas had no significant impact on the participant's preference for the color, hardness, brittleness, sweetness, and sourness of the dried apricots, while the participant preference for aroma and general acceptance were decreased at the concentrations of 28.6 and 33.4 mol% (Table 3).

Table 3. Mean panelists ranking scores of the sensory properties of the dried apricot treated with different concentration of CO_2.

Concentration (mol%)	Aroma	Color	Hardness	Brittleness	Sweetness	Sourness	General Acceptance
Control = 0	3.80 ± 0.20 [b]	5.10 ± 0.23 [a]	2.10 ± 0.18 [a]	2.40 ± 0.26 [a]	4.30 ± 0.15 [a]	3.30 ± 0.26 [a]	4.80 ± 0.29 [c]
9.1	3.70 ± 0.26 [b]	5.10 ± 0.23 [a]	2.00 ± 0.21 [a]	2.10 ± 0.31 [a]	4.40 ± 0.26 [a]	3.20 ± 1.35 [a]	4.90 ± 0.23 [c]
16.7	3.60 ± 0.22 [b]	4.90 ± 0.23 [a]	2.00 ± 0.21 [a]	2.00 ± 1.33 [a]	4.10 ± 1.37 [a]	3.60 ± 1.47 [a]	4.30 ± 0.26 [bc]
23.1	3.70 ± 0.21 [b]	4.90 ± 0.18 [a]	2.10 ± 0.21 [a]	1.90 ± 0.27 [a]	3.50 ± 1.45 [a]	3.60 ± 1.42 [a]	4.20 ± 0.29 [bc]
28.6	2.60 ± 0.26 [a]	4.70 ± 1.36 [a]	2.20 ± 0.21 [a]	1.90 ± 0.27 [a]	4.633 ± 0.56 [a]	3.50 ± 0.42 [a]	3.40 ± 0.34 [ab]
33.4	2.00 ± 0.25 [a]	4.90 ± 0.27 [a]	2.20 ± 0.13 [a]	2.20 ± 0.29 [a]	4.30 ± 0.36 [a]	3.60 ± 0.34 [a]	2.40 ± 0.26 [a]
Friedman's test:	χ^2 = 26.83 df = 5 p = 0.0000006	χ^2 = 2.22 df = 5 p = 0.817	χ^2 = 1.63 df = 5 p = 0.898	χ^2 = 1.87 df = 5 p = 0.867	χ^2 = 4.83 df = 5 p = 0.963	χ^2 = 1.67 df = 5 p = 0.894	χ^2 = 28.67 df = 5 p = 0.000002

Means (± standard error) with different letters within each column are statistically different according to the Friedman's test.

RGB pictures taken with the digital camera and colorimetric factors (L*, a*, and b*) of the dried apricot treated by CO_2 were shown in Figure 1. In the images taken, the color parameters (L*, a*, and b*) were calculated for all pixels of the samples. Then the mean values of each parameter were estimated for each image. The values of the color parameters L*, a*, and b* were calculated as 27.970, 23.119, and 36.397, respectively.

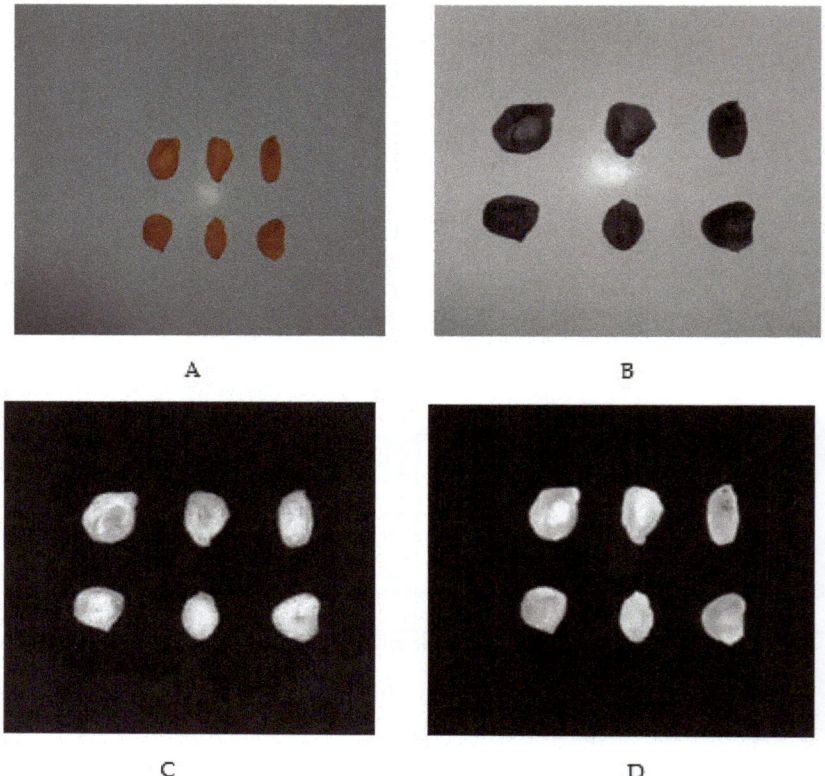

Figure 1. Image processing of the dried pieces of apricot: (**A**) RGB pictures taken with the digital camera; (**B**) Factor L* (brightness); (**C**) Factor a* (redness); (**D**) Factor b* (yellowness). In the images taken, the color parameters (L*, a*, and b*) were calculated for all pixels of the samples, and according to them, the mean values of each parameter were estimated (27.970, 23.119, and 36.397, respectively) through ImageJ 1.50 v software.

Figure 2 displays the increased changes of the brightness (ΔL*), redness (Δa*), and yellowness (Δb*) of the treated samples with enhancing CO_2 concentrations. Changes in the red color are almost constant at the concentrations between 16.7–23.1 and 28.6–33.4 mol%. Generally, alterations in the product color factors (L*, a*, and b*) induced by the CO_2 treatments are trivial. This degree of color changes triggered by CO_2 treatment is usually detectable with bare eyes.

Analysis of CO_2 gas impact at the concentration of 33.4 mol% on the chemical properties of the dried apricots, including pH, moisture content, acidity, Brix number, reducing sugar, and total sugar, and comparison with those of the control group revealed no significant difference between them (Table 4). Therefore, CO_2 gas at the concentration of 33.4 mol% would not affect the chemical properties of the control and treated dried pieces of apricot product.

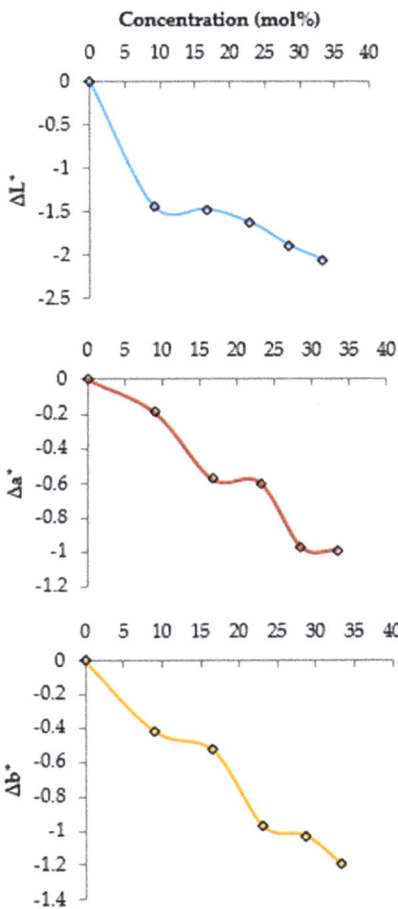

Figure 2. Effects of CO_2 concentrations on the color parameters (L* (brightness), a* (redness), and b* (yellowness)) of the dried apricots.

Table 4. Means ± standard errors of the chemical properties of the control and treated dried pieces of apricot after CO_2 gas treatment at the concentration of 33.4 mol%.

Chemical Properties	Control	Treatment	Comparison of Means (df = 4)	
			t	Sig.*
pH	4.17 ± 0.17	4.30 ± 0.08	−2.42	0.73
Acidity	1.13 ± 0.29	1.19 ± 0.27	−0.92	0.41
Brix	9.96 ± 0.44	9.82 ± 0.46	0.39	0.72
Total sugar	78.29 ± 3.07	78.79 ± 3.92	−0.14	0.9
Reducing sugar	56.86 ± 3.72	57.70 ± 4.67	−0.20	0.86
Moisture	10.30 ± 0.30	10.43 ± 1.70	−0.14	0.9

* Since the significant value is more than 0.05, tested means have no significant difference in each control and treatment row, according to the t test at $p < 0.05$. df and sig. are degree of freedom and significant, respectively.

4. Discussion

In current research, the mortality rates of two pest species were enhanced with increasing CO_2 concentration. The highest mortality, 78.0% and 90.5% for the adults of *T. castaneum* and *R. dominica*, respectively, was at the concentration of 33.4 mol%. Kells et al. [33] noticed the increasing CO_2 concentration from 25 to 50 ppm to enhance the larval mortality of Indian meal moth (*Plodia interpunctella* Hübner) from 77 to 94.5%. Riudavets et al. [34] reported that nine storage pests could be controlled by the CO_2 pressures of 15 and 20 bar within the time periods of 15, 30, and 60 min. In their study, 5.5% and 100% of the mortality of *R. dominica* adults were observed at the pressures of 15 and 20 bar after 15 min-exposure time, respectively. Accordingly, considering about 90% mortality by 0.5 bar pressure (33.4 mol%) of CO_2 in the present study, it will be possible to increase the mortality of *R. dominica* by increasing gas pressure (concentration) and exposure time. In research carried out by Sayeda and Hashem [35], the elevated mortality rate of Indian meal moth larvae from 38.9 to 77.8% was observed with increasing CO_2 concentration from 40 to 80% during 24 h of fumigation. These researchers reported CO_2 concentrations of 40% and 80% to lead to 35% and 71.7% larval mortality of almond moth (*Ephestia cautella* (Walker)), respectively. They experienced more vulnerabilities of the eggs and pupae of both species to the gas concentrations in comparison with larvae. At the concentration of 80%, 5 and 7 days required for full mortality of egg and pupal and larvae, respectively [35]. In the study of Wong-Corral et al. [36], increasing CO_2 concentration led to the enhanced mortality of Mexican bean weevil, *Zabrotes subfasciatus* (Boheman). After 2 days of treatment, gas concentration of 90% resulted in complete mortality, while the concentrations of 70 and 50% were required for complete mortality after 3 and 4 days of fumigation, respectively. However, they found that the adults of cowpea weevil and bean bruchid (*Acanthoscelides obtectus* (Say)) were more susceptible to the gas concentrations than Mexican bean weevil adults. In the research of Sadeghi et al. [37], the ozone concentration of 2 ppm and time period of 15 min resulted in the lowest mortality of the adults of the sawtoothed grain beetle (*Oryzaephilus surinamensis* L.) and larvae of flour moth *E. kuehniella*, while the concentration of 5 ppm and time of 90 min led to complete mortalities. These findings, about the susceptibility of stored-products insect pests to CO_2 gas, supports the results of current investigation. This is also congruent with our results, which showed increased mortality of the storage insects by elevating CO_2 gas concentrations.

In the present study, no significant impacts on the color, aroma, sweetness, sourness, brittleness, hardness, and general acceptance of the dried apricot were caused by enhancing the gas concentration of CO_2. Similarly, Sadeghi et al. [37] reported no significant influences of increasing ozone concentration and times of fumigation on the color, sweetness, sourness, brittleness, hardness, and general acceptance of dried figs and raisins, but they witnessed considerable effects on the aroma of the two products. Of course, they concluded that due to the volatility of ozone gas, their unpleasant odors could disappear over time. Accordingly, documented effects of CO_2 concentrations on the aroma of dried apricots can be removed over time. In the test of color changes of the product surface, the increases in CO_2 gas pressures were observed to have trivial effects on the surface colors of the dried apricot samples, which was not detectable with bare eye. In this regard, Sadeghi et al. [37,38] reported that increasing ozone concentrations (2, 3, and 5 ppm) and exposure times (15, 30, 45, 60, and 90 min) had no significant effects on the colorimetric factors of their raisins and figs.

Furthermore, CO_2 gas at the concentration of 33.4 mol% had no impacts on the chemical properties (pH, Brix value, acidity, reducing sugar, total sugar, and moisture content) of the dried apricots. These results were consistent with the findings of Inserra et al. [23], who reported insignificant impacts of sulfur on the moisture content, Brix value, and total sugar of dried apricot. The moisture percentage, pH, acidity, Brix value, reducing sugar, and total sugar contents of the dried apricots (Alkia cultivar) treated with sulfur were measured as be 22.5%, 4.2%, 1.8%, 68.1%, 64.4%, and 87.9%, respectively. These results, regarding no significant differences in chemical properties of treated and untreated dried apricot, are in agreement with those obtained in our study. The reason for the discrepancy of the measured Brix values in the study of Inserra et al. [23] and present study (68.1% and 9.82%, respectively) is related to the differences in the measurement methods, i.e., amounts of water added to the samples. In addition, the varied moisture contents can be attributed to the differences in the durations of the drying methods and their varieties [23].

5. Conclusions

According to the present results, fumigation of CO_2 caused significant toxicity on the adults of *R. dominica* and *T. castaneum* so that 20.19 and 27.02 mol% of the gas were adequate to kill 50% of pests within 24 h, respectively. The use of CO_2 gas, despite the toxicity to insect pests, had no significant side-effects on the sensory and chemical properties of the treated apricot specimens. Consequently, the prospective insecticidal activity of CO_2 gas offers an efficient tool to control the cosmopolitan insect pests *R. dominica* and *T. castaneum*, with no detrimental side-effects on stored-apricots. The main advantage of using carbon dioxide in preserving dried fruits such as apricots from pests is no necessity of complex and expensive devices. In other words, by using the cylinders containing this gas and adjusting the desired pressure, along with an impermeable covering of the warehouse, the control procedure can be carried out satisfactorily [39].

Author Contributions: Conceptualization, R.S., F.A. and F.H.; methodology, F.H.; formal analysis F.H. and A.E.; investigation, R.S., F.H. and A.E.; writing—original draft preparation, R.S., F.H., A.E., F.A., A.J. and F.P.; writing—review and editing, R.S., A.E. and F.P.; supervision, R.S., F.A. and A.J. All authors have read and agreed to the published version of the manuscript.

Funding: This research received no external funding.

Institutional Review Board Statement: Not applicable.

Informed Consent Statement: Not applicable.

Data Availability Statement: The data that support the findings of this study are available upon request from the authors.

Acknowledgments: The authors are grateful for financial support and valuable technical assistance to Department of Entomology and Plant Pathology, College of Aboureihan, University of Tehran. We acknowledge the Department of Food Technology, College of Aboureihan, University of Tehran for providing facilities to conduct this work.

Conflicts of Interest: The authors declare no conflict of interest.

References

1. Kan, T.; Gundogdu, M.; Ercisli, S.; Muradoglu, F.; Celik, F.; Gecer, M.K.; Kodad, O.; Zia-Ul-Haq, M. Phenolic compounds and vitamins in wild and cultivated apricot (*Prunus armeniaca* L.) fruits grown in irrigated and dry farming conditions. *Boil. Res.* **2014**, *47*, 46. [CrossRef]
2. Fellows, P.J. *Food Processing Technology, Principles and Practice*, 2nd ed.; Woodhead Publishing Limited: Cambridge, UK, 2000; pp. 378–405.
3. (FAO) Food and Agriculture Organization. *Countries by Commodity*; FAO: Rome, Italy, 2018.
4. Burks, S.C.; Johnson, J.A. Biology, Behavior, and Ecology of Stored Fruit and Nut Insects. In *Stored Product Protection*, 1st ed.; Hagstrum, D.W., Phillips, T.W., Cuperus, G., Eds.; Kansas State Research and Extension: Kansas, TX, USA, 2012; pp. 21–33.
5. Moreda, G.P.; Ruiz-Altisent, M. Quality of agricultural products in relation to physical conditions. In *Encyclopedia of Agrophysics*, 1st ed.; Gliński, J., Horabik, J., Lipiec, J., Eds.; Springer: Dordrecht, The Netherlands, 2011; pp. 669–678. [CrossRef]
6. Heckel, D.G. Insecticide resistance after silent spring. *Science* **2012**, *337*, 1612–1614. [CrossRef]
7. Nicolopoulou-Stamati, P.; Maipas, S.; Kotampasi, C.; Stamatis, P.; Hens, L. Chemical pesticides and human health: The urgent need for a new concept in agriculture. *Front. Public Health* **2016**, *4*, 148. [CrossRef] [PubMed]
8. Zikankuba, V.L.; Mwanyika, G.; Ntwenya, E.; James, A. Pesticide regulations and their malpractice implications on food and environment safety. *Cogent Food Agric.* **2019**, *5*, 1601544. [CrossRef]
9. Panna, K.S.S. Methyl Bromide Phase-Out Strategies; a Global Compilation of Laws and Regulations. 1999. Available online: http://www.unep.fr/ozonaction/information/mmcfiles/3020-e.pdf (accessed on 24 May 2021).
10. Damalas, C.A.; Eleftherohorinos, I.G. Pesticide exposure, safety issues, and risk assessment indicators. *Int. J. Environ. Res. Public Health* **2011**, *8*, 1402–1419. [CrossRef]

11. Ebadollahi, A.; Ziaee, M.; Palla, F. 2020: Essential oils extracted from different species of the Lamiaceae plant family as prospective bioagents against several detrimental pests. *Molecules* **2020**, *25*, 1556. [CrossRef]
12. Palla, F.; Bruno, M.; Mercurio, F.; Tantillo, A.; Rotolo, V. Essential oil as natural biocides in conservation of cultural heritage. *Molecules* **2020**, *25*, 730. [CrossRef] [PubMed]
13. Cao, Y.; Xu, K.; Zhu, X.; Bai, Y.; Yang, W.; Li, C. Role of modified atmosphere in pest control and mechanism of its effect on insects. *Front. Physiol.* **2019**, *10*, 206. [CrossRef]
14. Hasan, M.M.; Aikins, M.J.; Schilling, W.; Phillips, T.W. Efficacy of controlled atmosphere treatments to manage arthropod pests of dry-cured hams. *Insects* **2016**, *7*, 44. [CrossRef]
15. Khalil, S.S.H.; Ahmed, S.S.; Abd El-Rahman, S.F.; Khalifa, E.A.A. Comparative effects of carbon dioxide concentration in moified temperature and pressure conditions on adults and larvae of the red flour beetle, *Tribolium castaneum* (Herbst) (Coleoptera: Tenebrionidae). *Coleop. Bull.* **2020**, *74*, 127–135. [CrossRef]
16. Leelaja, B.C.; Rajashekar, Y.; Reddy, P.V.; Begum, K.; Rajendran, S. Enhanced fumigant toxicity of allyl acetate to stored-product beetles in the presence of carbon dioxide. *J. Stored Prod. Res.* **2006**, *43*, 45–48. [CrossRef]
17. Cheng, W.; Lei, J.; Ahn, J.E.; Liu, T.X.; Zhu-Salzman, K. Effects of decreased O_2 and elevated CO_2 on survival, development, and gene expression in cowpea Bruchids. *J. Insect Physiol.* **2012**, *58*, 792–800. [CrossRef] [PubMed]
18. Valizadegan, O.; Pourmirza, A.A.; Safaralizadeh, M.H. The impact of carbon dioxide in stored-product insect treatment with phosphine. *Afr. J. Biotechnol.* **2012**, *11*, 6377–6382. [CrossRef]
19. Dhouibi, M.H.; Hermi, N.; Hajjem, B.; Hammami, Y.; Arfa, M.B. Efficacy of CO and CO_2 treatment to control the stored-date insects, *Ephestia kuehniella* and *Ectomyelois ceratoniae* for organic dates production. *Int. J. Agric. Innov. Res.* **2015**, *3*, 1473–2319.
20. Sun, D.W.; Brosnan, T. Inspection and grading of agricultural and food products by computer vision systems-a review. *J. Comput. Electron. Agric.* **2002**, *36*, 193–213. [CrossRef]
21. Leon, K.; Mery, D.; Pedreschi, F.; Leon, J. Color measurement in L* a* b* units from RGB digital images. *J. Food Int. Res.* **2006**, *39*, 1084–1091. [CrossRef]
22. Sadeghi, R.; Seyedabadi, A.; Mirabi Moghaddam, R. Microwave application for controlling *Oryzaephilus surinamensis* Insects infesting dried figs and evaluation of product color changes using an image processing technique. *J. Food Prot.* **2018**, *82*, 184–188. [CrossRef]
23. Inserra, L.; Cabaroglu, T.; Sen., K.; Arena, E.; Ballistreri, G.; Fallico, B. Effect of sulphuring on physicochemical characteistics and aroma of dried Alkaya apricot: A new Turkish variety. *J. Agric. For.* **2017**, *41*, 59–68. [CrossRef]
24. Rees, D. *Insects of Stored Grain: A Pocket Reference*, 2nd ed.; CSIRO Publishing: Collingwood, Australia, 2007; p. 77.
25. Sarwar, M. Protecting dried fruits and vegetables against insect pests invasions during drying and storage. *Am. J. Market. Res.* **2015**, *1*, 142–149.
26. Sousa, A.H.; Faroni, L.R.D.A.; Guedes, R.N.C.; Totola, M.R.; Urruchi, W.I. Ozone as a management alternative against phosphine-resistant insect pests of stored products. *J. Stored Prod. Res.* **2008**, *44*, 379–385. [CrossRef]
27. Vayias, B.J.; Athanassiou, C.G.; Kavallieratos, N.G.; Tsesmeli, C.D.; Buchelos, C.T. Persistence and efficacy of two diatomaceous earth formulations and a mixture of diatomaceous earth with natural pyrethrum against *Tribolium confusum* Jacquelin du Val (Coleoptera: Tenebrionidae) on wheat and maize. *Pest Manag. Sci.* **2006**, *62*, 456–464. [CrossRef]
28. Subramanyam, B.; Toews, M.D.; Ileleji, K.E.; Maier, D.E.; Thompson, G.D.; Pitts, T.J. Evaluation of spinosad as a grain protectant on three Kansas farms. *Crop Prot.* **2007**, *26*, 1021–1030. [CrossRef]
29. (INSO) Iranian National Standards Organization. *Sensory Evaluation-Methodology-Tips for Monitoring the Performance of a Quantitative Sensory Evaluation Team, Standard No. 18294*; NISO: Tehran, Iran, 2014.
30. El-Arem, A.; Guido, F.; Behija, S.E.; Manel, I.; Nesrine, I.; Ali, Z.F.; Mohamed, H.; Noureddine, H.; Lotfi, A. Chemical and aroma volatile compositions of date palm (*Phoenix dactylifera* L.) fruits at three maturation stages. *J. Food Chem.* **2011**, *127*, 1744–1754. [CrossRef]
31. Parvaneh, V. *Quality Control of Food Chemical Experiments*, 7th ed.; University of Tehran Publications: Tehran, Iran, 2013; pp. 210–247.
32. Abbott, W.S. A method of computing the effectiveness of an insecticide. *J. Econ. Entomol.* **1925**, *18*, 265–267. [CrossRef]
33. Kells, S.A.; Mason, L.J.; Maier, D.E.; Woloshuk, C.P. Efficacy and fumigation characteristics of ozone in stored maize. *J. Stored Prod. Res.* **2001**, *37*, 371–382. [CrossRef]
34. Riudavets, J.; Castañé, C.; Alomar, O.; Pons, M.J.; Gabarra, R. The use of carbon dioxide at high pressure to control nine stored-product pests. *J. Stored Prod. Res.* **2010**, *46*, 228–233. [CrossRef]
35. Sayeda, A.; Mohammad, Y.H. Susceptibility of different life stages of Indian meal moth *Plodia interpunctella* (Hübner) and almond moth *Ephestia cautella* (Walker) (Lepidoptera: Pyralidae) to modified atmospheres enriched with carbon dioxide. *J. Stored Prod. Res.* **2012**, *51*, 49–55. [CrossRef]
36. Wong-Corral, F.J.; Castañé, C.; Riudavates, J. Lethal effects of CO_2-modified atmosphere for the control of three Bruchidae species. *J. Stored Prod. Res.* **2013**, *55*, 62–67. [CrossRef]
37. Sadeghi, R.; Mirabi Moghaddam, R.; Taghizadeh, M. Application of ozone to control dried fig pests *Oryzaephilus surinamensis* (Coleoptera: Silvanidae) and *Ephestia kuehniella* (Lepidoptera: Pyralidae) and its organoleptic properties. *J. Econ. Entomol.* **2017**, *110*, 2052–2055. [CrossRef]

38. Sadeghi, R.; Seyedabadi, A.; Mirabi Moghaddam, R. Evaluation of microwave and ozone disinfections on the color characteristics of Iranian export raisins through an image processing technique. *J. Food Prot.* **2019**, *82*, 2080–2087. [CrossRef]
39. Navarro, S. The use of modified and controlled atmospheres for the disinfestation of stored products. *J. Pest Sci.* **2012**, *85*, 301–322. [CrossRef]

Article

Effect of Cold Smoking and Natural Antioxidants on Quality Traits, Safety and Shelf Life of Farmed Meagre (*Argyrosomus regius*) Fillets, as a Strategy to Diversify Aquaculture Products

Concetta Maria Messina [1,2,*], Rosaria Arena [1], Giovanna Ficano [1], Mariano Randazzo [2], Maria Morghese [1], Laura La Barbera [2], Saloua Sadok [3] and Andrea Santulli [1,2]

Citation: Messina, C.M.; Arena, R.; Ficano, G.; Randazzo, M.; Morghese, M.; La Barbera, L.; Sadok, S.; Santulli, A. Effect of Cold Smoking and Natural Antioxidants on Quality Traits, Safety and Shelf Life of Farmed Meagre (*Argyrosomus regius*) Fillets, as a Strategy to Diversify Aquaculture Products. *Foods* **2021**, *10*, 2522. https://doi.org/10.3390/foods10112522

Academic Editors: Lisa Pilkington and Siew-Young Quek

Received: 12 September 2021
Accepted: 16 October 2021
Published: 21 October 2021

Publisher's Note: MDPI stays neutral with regard to jurisdictional claims in published maps and institutional affiliations.

Copyright: © 2021 by the authors. Licensee MDPI, Basel, Switzerland. This article is an open access article distributed under the terms and conditions of the Creative Commons Attribution (CC BY) license (https://creativecommons.org/licenses/by/4.0/).

[1] Dipartimento di Scienze della Terra e del Mare DiSTeM, Laboratorio di Biochimica Marina ed Ecotossicologia, Università degli Studi di Palermo, Via G. Barlotta 4, 91100 Trapani, Italy; rosaria.arena@unipa.it (R.A.); giovanna.ficano@unipa.it (G.F.); maria.morghese@unipa.it (M.M.); andrea.santulli@unipa.it (A.S.)

[2] Istituto di Biologia Marina, Consorzio Universitario della Provincia di Trapani, Via G. Barlotta 4, 91100 Trapani, Italy; mariano.randazzo@tin.it (M.R.); labarbera@consunitp.it (L.L.B.)

[3] Laboratory of Blue Biotechnology & Aquatic Bioproducts (B3Aqua), Institut National des Sciences et Technologies de la Mer (INSTM), Annexe La Goulette Port de Pêche, La Goulette 2060, Tunisia; salwa.sadok@instm.rnrt.tn

* Correspondence: concetta.messina@unipa.it

Abstract: Aquaculture has been playing a leading role over the years to satisfy the global growing demand for seafood. Moreover, innovative techniques are necessary to increase the competitiveness, sustainability and profitability of the seafood production chain, exploiting new species from the aquaculture, such as meagre (*Argyrosomus regius*), to develop value-added products and diversify their production. In the present work, the effectiveness of cold smoking combined with antioxidants (SA) compared to cold smoking alone (S) on meagre fillets, the quality and shelf life were investigated. Sensory, biochemical, physical–chemical and microbiological analyses were performed on the smoked fillets during vacuum-packaged storage for 35 days at 4 ± 0.5 °C. The results showed positive effects of the SA treatment on the biochemical parameters of meagre fillets. The total volatile basic nitrogen (TVB-N) in smoked meagre fillets was significantly lower in the SA treatment at the end of storage compared to the S treatment. Moreover, SA had a positive effect on lipid peroxidation. Lower values of malondialdehyde (mg MDA/kg) were observed in the SA treatment during preservation compared to the S treatment. This work will contribute to the growth of the fish production chain, producing a value-added fish product by exploiting meagre, whose production has been increasing over decades.

Keywords: meagre; *Argyrosomus regius*; cold smoking; natural antioxidants; halophyte; aquaculture; value-added food product; shelf-life; fillets; fish quality

1. Introduction

Global fish consumption has been increasing over the years due to the increasing world population, as well as the awareness of the beneficial effects of fish inclusion in a balanced diet on human health. It has been reported that the average annual rate at which global fish consumption increased from 1961 to 2017 accounted for 3.1%, almost twice that of the annual growth of the global population [1].

In this context, the fish production sector has grown over the years in order to satisfy the global growing demand for seafood, and it is foreseen to increase in the future. Specifically, the production of capture fisheries has not been increasing since the late 1980s, while aquaculture has been playing a leading role in satisfying the seafood demand over the years. In this scenario, the aquaculture contribution to fish production has constantly increased at the global level, reaching 46.0% in the period 2016–2018. In 2018, aquaculture accounted for 52% of the total production of seafood products that were used for human consumption. Specifically, in 2018, aquaculture produced 82.5 million tons of fish and other organisms, 32.4 million tons of plant organisms and a good amount of products not intended for consumption (26,000 tons) [1].

Innovation plays the major role in increasing competitiveness, sustainability and profitability of the seafood production chain, paying attention to develop value-added processed products by exploiting new species coming from aquaculture, an ever-increasing food production sector. This strategy would allow to satisfy consumers' preferences in the food sector, who prefer ready-to-eat products with high nutritional value whose availability does not determine a negative effect on the environment and whose origin and quality are guaranteed [2]. In this context, meagre (*Argyrosomus regius*) represents an aquaculture species that must be considered and valorized, considering that its production in Europe has been increasing over the years. Specifically, meagre production has been rising progressively in Europe over the decades; in 2018, the production of meagre reached 6.827 tons (+270%), out of which 59% was covered by Spain [3].

Meagre, which belongs to the *Sciaenidae*, is a euryhaline fish spread in the Mediterranean Sea. Thanks to some characteristics of *A. regius*, such as its high adaptability to several environmental conditions and high resilience to stressors, it has been considered suitable for the diversification of the Mediterranean aquaculture [4]. In addition to these aspects, *A. regius* grows rapidly and reaches commercial size (700–1200 g) only after 12 months and 2–2.5 kg after 24 months under optimal conditions [4]. Moreover, meagre is characterized by a high nutritional value thanks to its flesh, lean and with a high lipid quality [5].

Fish is the main dietary source of omega-3 polyunsaturated fatty acids (PUFAs), highly beneficial for human health but also extremely prone to peroxidation, such as docosahexaenoic acid (DHA) and eicosapentaenoic acid (EPA). In particular, their oxidation generates free radicals that trigger chain reactions that eventually lead to the generation of compounds responsible for extremely unpleasant odors, as well as the loss of compounds of very high nutritional value [6,7]. Considering the characteristics of the raw material processed, highly perishable, preservatives should be used to preserve the fish quality and extend the shelf life; indeed, even if fish is preserved by freezing or cold storage to extend its shelf life, this would not be sufficient to prevent fish spoilage, lipid oxidation and rancidity with negative effects in terms of consumer acceptability.

Chemical preservatives have been used in order to prevent fish spoilage and extend the fish shelf life, such as sodium acetate, sodium lactate and sodium citrate, whose effects were investigated on extending the shelf life of salmon (*Onchorhynchus nerka*) during refrigerated storage [8], highlighting that using these preservatives was effective on extending seafood shelf life. Sodium bisulfite is also used as a chemical preservative to reduce enzyme activity and microbial degradation in shrimp (*Litopenaeus vannamei*) for 4–6 days; nevertheless, it has been reported that consuming food treated with sodium bisulfite for a long time can lead to a serious health crisis [9].

Therefore, the food industry has been prompt about consumers' concerns and preferences for natural preservatives for the use of natural antioxidants and antimicrobials instead of synthetic ones [7]. Marine and terrestrial species have thus been used for the purpose of extracting antioxidant compounds, such as ascorbic acid, glutathione, carotenoids, terpenes and phenolic compounds [7,10–14].

Among plant-derived compounds, there are essential oils, a complex mixture of volatile compounds produced as secondary metabolites in plants and that give them their characteristic odor. Essential oils have been used as natural preservatives for preserving fish quality and extending their shelf life due to their well-recognized antimicrobial and antioxidants potentials [15] Plant extracts could also been used for application in fish preservation for their antimicrobial and antifungal activities, as well as their capacity to inhibit lipid oxidation [7,16].

Salicornia strobilacea (*Halocnemum strobilaceum*) is a plant that belongs to halophytes and grows along salt marshes, salt lakes and saltworks all over the world. Due to the extreme environmental conditions under which this kind of plant grows, it is characterized by powerful antioxidant systems that employ several components, among which are the secretion and accumulation of polyphenols. These bioactive compounds were obtained from the plant (*H. strobilaceum*) and tested in vitro, showing strong antioxidant and antibacterial power [17].

The positive effects of antioxidants from *H. strobilaceum* when combined with other preservative techniques such as modified atmosphere packaging and cold smoking were shown previously on the sensory, physical–chemical, nutritional, biochemical and microbiological properties of dolphinfish (*Coryphaena hippurus*) fillets [18,19].

Cold smoking is a preservation technique also used to create new value-added products. Specifically, in cold smoking, after an initial treatment with salt, the product is smoked in order to give the product organoleptic characteristics appreciated by consumers, as well as to transfer to it antimicrobial and antioxidant compounds (aldehydes, ketones, alcohols and phenols), with a significantly longer shelf-life than fresh product. During the different phases of the process, the temperature is never higher than 30 or 33 °C [20,21]. To date, several fish species have been processed by this technique, such as farmed European sea bass (*Dicentrarchus labrax*) and Atlantic salmon (*Salmo salar*) [22,23]. Moreover, fishery species have been valorized by applying cold smoking in order to make available all year round seasonal species, as in the case of dolphinfish (*Coryphaena hippurus*), but also to enhance fishery species with excess catches, such as sardines (*Sardina pilchardus*) [19,24,25] and herring (*Clupea harengus*) [26].

The effectiveness of combining traditional techniques such as cold-smoking with the addition of natural preservatives on maintaining the quality and prolonging the shelf life of new value-added seafood products has been investigated [19,27]. Combining an antioxidant treatment with oregano extract and cold-smoking produced a reduction in sardine (*S. pilchardus*) lipid oxidation compared to traditional smoking, highlighting the effectiveness of the combined treatments on improving the shelf life of smoked fish [27].

The combination of cold smoking and natural preservatives was also investigated on dolphinfish (*C. hippurus*) fillets [19]. Overall, the combination of antioxidants with cold smoking showed positive effects on the quality of dolphinfish fillets, improving the biochemical, microbiological and sensory aspects of the product and, consequently, enhancing the marketability of *C. hippurus* and contributing to costal fishery sustainability.

The present work was aimed at studying how the combination of cold smoking and natural antioxidants affects the sensory, biochemical, physical–chemical and microbiological properties of meagre (*Argyrosomus regius*) fillets. As a source of antioxidants, *H. strobilaceum* extracts were used.

The present study will contribute to increasing the competitiveness, profitability and sustainability of the seafood production sector by developing a new value-added product by using an aquaculture species (*A. regius*) of high nutritional value whose production has been increasing over the years, contributing to diversifying seafood production, meeting consumers preferences and reducing the pressure on overexploited fishery resources.

2. Materials and Methods

2.1. Fish Sampling and Processing

A total of 20 specimens of *A. regius* (average size 27.17 ± 1.91 cm and average weight 347.72 ± 62.85 g) were processed in a Sicilian (Italy) aquaculture plant to obtain forty fillets (mean weight: 73.48 ± 22.49 g) that were stored on ice. In the laboratory (less than 30 min), the fillets were stored under vacuum in Foodsaver bags (HDPE) and nylon bags (http://www.gopack.it, accessed on 1 September 2021) and subjected to rapid freezing at −35 °C (AB 2/3 ALLFORFOODD (PU), Italy) and maintained in this condition for 24 h to inhibit the bacterial growth [28].

The next day, all samples were removed from the bags, placed in air-permeable LDPE bags and thawed at 4 °C for 8 h before processing.

2.2. Salting and Smoking

Thawed *A. regius* fillets were processed as follows: four thawed fillets were used for the analyses of the untreated product, and the remaining 36 fillets underwent a smoking process consisting of four steps: salting, first drying, smoking and second drying [19,22], as shown in Figure 1.

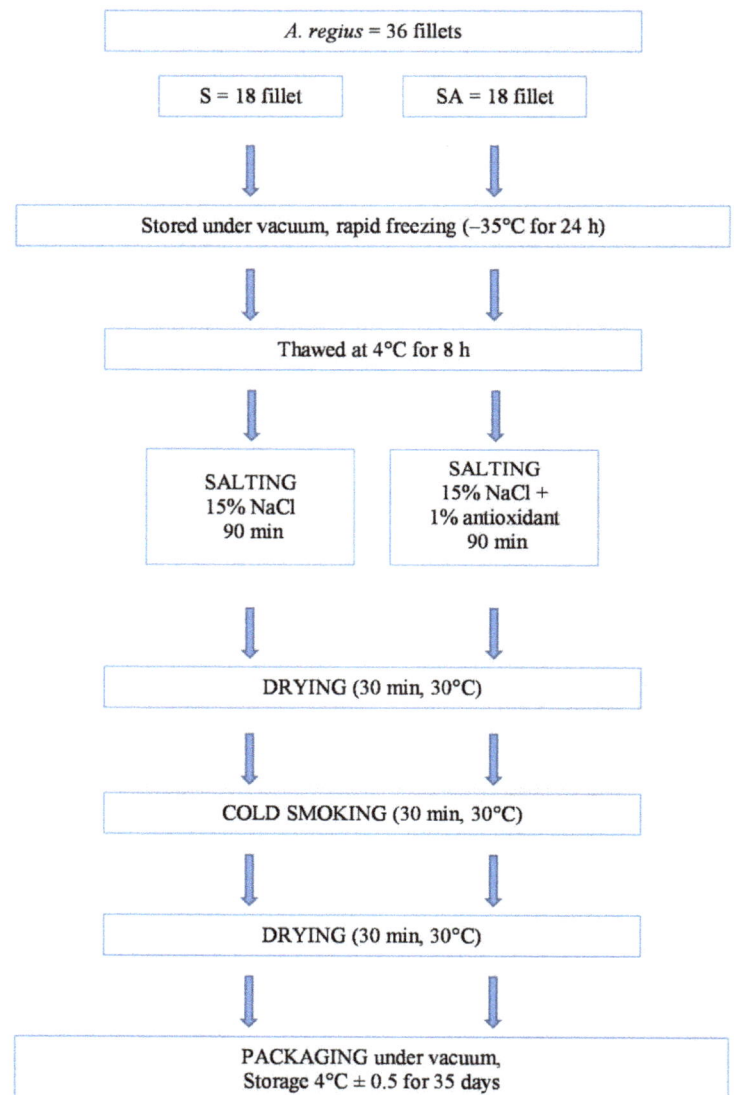

Figure 1. Experimental design flowchart. Cold smoking (S); Cold smoking combined with antioxidants (SA).

Thawed fillets were separated in 2 batches: in the first batch, 18 fillets were immersed in standard brine (S) consisting of a 15% NaCl solution (w/v); in the second batch, 18 fillets were treated with the same standard brine with 1% antioxidants (SA). For both treatments, a fillet:brine ratio of 1:4 was used [19,22].

The antioxidant solution was prepared as described by Messina et al. [18]. Briefly, dried and pulverized *Halocnemum strobilaceum* was extracted with distilled water (1:10 *w/v*) for 24 h. The sample was then filtered and lyophilized [29]. The final solution of *H. strobilaceum* was prepared by dissolving 10 g of freeze-dried extract in 1000 mL of distilled water, with a polyphenol content equal to 500-mg gallic acid equivalent (GAE)/L [18,19].

Brine salting was performed for 90 min; then, the samples were dried for 30 min at a temperature of 30 °C. The fillets were cold-smoked using Moduline oven model FA082E (Scubla srl, Remanzacco (Ud), Italy) for 30 min at 30 °C, as described by Messina et al. [19,22].

After cold smoking, the fillets were dried as described previously.

At the end of the process, all the fillets were sealed in vacuum bags and stored at 4 ± 0.5 °C for 35 days.

Three fillets from each treatment were analyzed at regular intervals (1, 7, 14, 21, 28 and 35 days after smoking). The effects of the smoking process combined with natural antioxidants on the quality of *A. regius* fillets were evaluated through a multidisciplinary approach involving sensory, physicochemical, biochemical and microbiological parameters.

Following sampling for the microbiological analysis and the sensory and instrumental analyses, the fillets were cold-homogenized for the analysis of the biochemical and nutritional parameters related to shelf life.

2.3. Physical–Chemical Parameters
2.3.1. Color

Color readings were taken in the L*, a* and b* color space (CIELAB color space, D65 standard illumination and a 2° observer) and repeated 3 times using a Konica Minolta colorimeter (Osaka, Japan). The evaluated color parameters were lightness (L*), red–green chromaticity (a*), yellow–blue chromaticity (b*), the saturation or intensity of the color chroma (C*) and the hue angle (h), as recommended by the International Commission on Illumination [30–32]. The analyses were performed in triplicate for each sample. The color evaluation was carried out in two dorsal regions of the fillets along the cephalocaudal direction.

2.3.2. Texture Profile Analysis

The texture analysis was conducted as described by Messina et al. [33]. Two small fragments (1.8 cm Ø) obtained from the same portions of each fillet were used. The analysis was performed at room temperature using Instron Texture Analyzer Mod. 3342 (Turin, Italy). The measured parameters were hardness (N) and the Young's modulus or modulus of deformability (N/mm^2) (i.e., the force and the slope of the curve at 50% compression, respectively) [19,34]. The analysis was performed in triplicate. In particular, for each replicate mentioned above, two fragments per fillet along the dorsal margin were considered. The samples were maintained in ice before the analysis.

2.3.3. Water Holding Capacity (WHC)

The Water holding capacity (WHC) was determined using the method described by Teixeira et al. [35], with some modifications reported by Messina et al. [33]. The analyses were performed in triplicate for each sample. The results were expressed in percentage (% WHC).

2.3.4. Muscular pH

The muscular pH of the fillet was measured at three points along the lateral line with a Crison pH meter (Barcelona, Spain) equipped with a BlueLine, pH 21 Schott Instruments (Weilheim, Germany) combined electrode.

2.3.5. Water Activity Determination

The water activity (aw) was measured with a fast water activity meter (HP23-AW Rotronic, AG, Bassersdorf, Switzerland). The temperature at which the water activity was measured was equal to 21 °C. The analysis was performed in triplicate.

2.4. Proximate Composition and Biochemical Parameters Related to the Shelf-Life

2.4.1. Proximate Composition

After the physical–chemical and sensory analyses, in order to carry out further analyses, the fillets were cold-homogenized.

The ash (ignition at 600 °C for 5 h) and moisture (drying at 105 °C for 24 h) contents (% ash and moisture) were assessed according to the AOAC method [36]. The protein content (% protein) was determined according to the AOAC method [37].

The total lipids (% lipids) were determined according to Folch et al. [38], and the fatty acid (FA) methyl esters (%) were determined by the method of Lepage and Roy [39]; gas-chromatography was carried out following the operating conditions described by Messina et al. [40].

2.4.2. Biochemical Parameters Related to the Shelf-Life

The production of thiobarbituric acid reactive substances (TBARS) was determined using the method described by Botsoglou et al. [41] and the results expressed in mg MDA (malondialdehyde)/kg. The total volatile basic nitrogen (TVB-N) was measured by direct distillation of the homogenized samples according to the EU Commission Decision 95/149/EC [42] and the values expressed in mg/100 g of product. All analyses were performed in triplicate.

2.5. Microbiological Analyses

The microbiological analysis was performed as described by Messina et al. [19].

2.6. Sensorial Analysis

A sensory analysis on cold-smoked meagre fillets was conducted by a group of six trained judges according to an adapted version of the scheme proposed by Bilgin et al. [43] on hot-smoked meagre.

The parameters evaluated were appearance, odor, flavor, texture and color. The six judges rated the overall acceptability of the samples using a 10-point descriptive scale. A score of 9 to 10 indicates a perfect product, 7 to 8 good, 5 to 6 medium and 3 to 4 the limit of acceptability. The product is considered unacceptable when scoring less than 3 [43].

2.7. Statistical Analysis

The results were expressed as the mean ± standard deviation. Homogeneity of the variance was analyzed by Levene's test. The data were analyzed by one-way analysis of variance (ANOVA), and Student–Newman–Keuls and Games–Howell post hoc tests were performed to make multiple comparisons between the experimental groups. Differences were statistically significant when $p < 0.05$. All data were analyzed by the SPSS for Windows® application (version 15.0, SPSS, Chicago, IL, USA).

3. Results and Discussion

3.1. Physical–Chemical Parameters

Smoking is a process that alters the physical and chemical properties of the raw material, such as color, texture, pH, aw, water holding capacity (WHC), etc. It is important to control the physical and chemical parameters in the processed product, as they affect the shelf life and sensory characteristics [44].

3.1.1. Color

Color is an important attribute of fish quality and freshness and also indicative of the chemical components and sensory attributes of food [45].

The results of the instrumental color analysis, in terms of L*, a* and b*; C* and h coordinates, obtained from *A. regius* fillets, are showed in Table 1.

Table 1. Results of the color of cold-smoked meagre samples during storage at 4 ± 1 °C.

	Storage Time (Days)	S	SA
L* (D65)	Thawed	45.48 ± 1.43 [c]	44.65 ± 1.41 [d]
	0	40.86 ± 1.26 [a]	39.87 ± 1.13 [b]
	1	40.52 ± 0.57 [a]	39.46 ± 0.18 [b]
	7	39.17 ± 0.67 [a]	38.87 ± 0.10 [ab]
	14	40.96 ± 1.03 [a/B]	37.96 ± 0.41 [a/A]
	21	40.14 ± 0.18 [a/B]	38.36 ± 0.07 [a/A]
	28	39.90 ± 0.76 [a]	39.84 ± 0.59 [b]
	35	42.72 ± 0.33 [b/B]	41.38 ± 0.39 [c/A]
a* (D65)	Thawed	-2.12 ± 0.39 [d]	-2.31 ± 0.43 [c]
	0	-2.35 ± 0.28 [cd]	-2.72 ± 0.31 [bc]
	1	-2.57 ± 0.07 [c/B]	-2.85 ± 0.10 [bc/A]
	7	-3.47 ± 0.36 [b]	-3.55 ± 0.12 [ab]
	14	-3.29 ± 0.03 [b]	-2.63 ± 0.79 [b]
	21	-4.31 ± 0.02 [a/A]	-3.50 ± 0.10 [ab/B]
	28	-3.61 ± 0.06 [b/A]	-3.09 ± 0.09 [b/B]
	35	-3.31 ± 0.39 [b/B]	-4.25 ± 0.10 [a/A]
b* (D65)	Thawed	-3.56 ± 0.83 [b]	-2.61 ± 0.75 [a]
	0	-6.67 ± 0.59 [a/A]	-3.90 ± 0.68 [a/B]
	1	-5.62 ± 0.63 [b/A]	-2.80 ± 0.58 [a/B]
	7	-3.41 ± 0.96 [c/A]	-1.05 ± 0.01 [b/B]
	14	-2.20 ± 0.13 [cd/A]	0.37 ± 0.37 [c/B]
	21	-0.75 ± 0.64 [d]	0.43 ± 0.22 [c]
	28	-0.14 ± 0.54 [d]	1.41 ± 0.59 [c]
	35	-1.68 ± 0.79 [cd/A]	1.51 ± 0.08 [c/B]
C* (D65)	Thawed	4.21 ± 0.84 [a]	3.53 ± 0.82 [ab]
	0	7.09 ± 0.62 [b/B]	4.78 ± 0.69 [b/A]
	1	6.25 ± 0.53 [b/B]	4.04 ± 0.47 [ab/A]
	7	4.91 ± 0.92 [a]	3.82 ± 0.05 [ab]
	14	4.16 ± 0.23 [a]	2.81 ± 0.67 [a]
	21	4.53 ± 0.16 [a/B]	3.68 ± 0.16 [ab/A]
	28	3.80 ± 0.01 [a]	3.50 ± 0.31 [ab]
	35	3.93 ± 0.01 [a/A]	4.60 ± 0.14 [b/B]
h (D65)	Thawed	238.17 ± 4.55 [d]	226.24 ± 5.82 [cd]
	0	250.57 ± 1.60 [e/B]	234.51 ± 3.37 [d/A]
	1	244.05 ± 2.92 [d/B]	222.83 ± 5.22 [c/A]
	7	222.51 ± 5.36 [c/B]	195.21 ± 0.43 [b/A]
	14	211.64 ± 0.05 [bc/B]	167.86 ± 10.09 [a/A]
	21	188.24 ± 7.93 [a]	172.47 ± 3.32 [a]
	28	180.02 ± 8.95 [ab]	156.66 ± 8.67 [a]
	35	206.35 ± 13.97 [bc/B]	161.40 ± 0.16 [a/A]

The means with different letters (a, b and c) in the same column are significantly different ($p < 0.05$). The means with different letters (A, B and C) in the row are significantly different ($p < 0.05$).

As for the L* parameter (Table 1), it was observed that the cold-smoking process significantly affected the lightness, as both S and SA fillets showed a significant decrease ($p < 0.05$) of this parameter immediately after smoking. This is due to the effect of smoking, which induces a loss of water, which could have led to an increase in the carotenoid concentration and a decrease in hue and lightness [22,46–48].

During the shelf life, the L* parameter remained constant and increased significantly ($p < 0.5$) in the last days of storage (Table 1). This significant increase was also observed in other species such as sea bass [33].

Significant differences ($p < 0.05$) were observed between the two treatments (S and SA) starting from the 14th day of storage; the S samples showed significantly higher L* values than the SA samples, probably due to a higher moisture content in the sample. In fact, the higher percentage of water contributes to the creation of refractive indices in the food matrix, which leads to a greater luminosity [49,50].

The parameter a* decreased significantly ($p < 0.05$) during the shelf life in both treatments (S and SA). This decreased redness, mainly caused by the smoking process, has also been reported in salmon [51]; in fact, a general tendency has been observed in cold-smoked salmon to be darker and less red [51].

Regarding parameter b*, it decreased significantly after smoking ($p < 0.05$), then increased during storage (Table 1).

This yellowness of the muscle may be due to the resulting darkening of the bloodline due to hemeprotein oxidation, associated with a corresponding reduction in a* [22,52].

As shown in the present study, a general trend in cold-smoked meagre fillets was observed, i.e., they were darker and less red but more yellowish than unprocessed fillets. This yellowness was more evident in fillets treated with the natural antioxidant added during salting. Probably the coloration of the antioxidant affected the coloration of the final product.

3.1.2. Texture Profile Analysis

The synergistic action of salt incorporation, the preservative effect of smoke compounds and dehydration during the smoking process can preserve fish. This leads to an increase in the tissue parameters, such as hardness [19,22,48,53–57].

The results of the texture analysis (Young's modulus and hardness) are shown in Figure 2.

Young's modulus (or modulus of deformability) values (Figure 2a) increased significantly ($p < 0.05$) after cold smoking and remained constant until T 28; a significant decrease ($p < 0.05$) was observed at T 35. No significant differences were observed between the two treatments.

The hardness value (Figure 2b) increased significantly until the 14th day of storage (Table 1), but no significant differences were observed between the treatments.

The increase in hardness was due to the synergistic effect of salting and drying causing meat hardening [53].

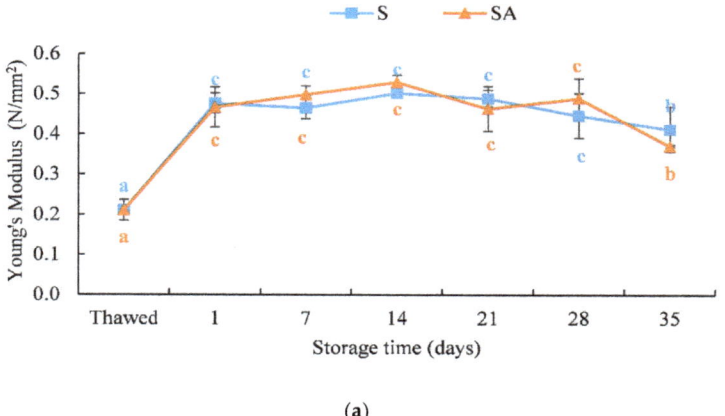

(a)

Figure 2. *Cont.*

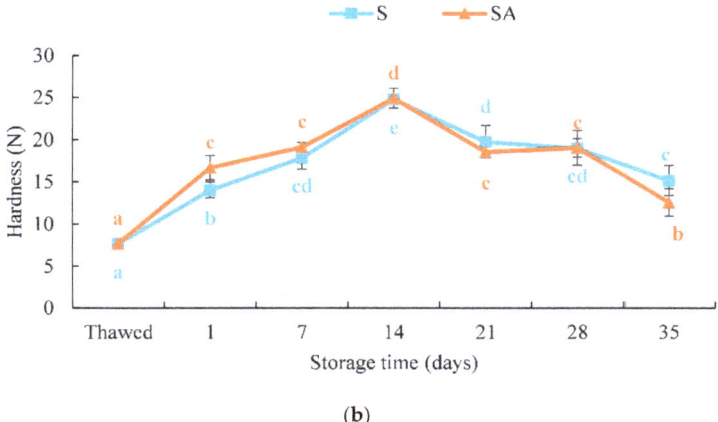

(b)

Figure 2. Results of the texture of cold-smoked meagre samples during storage at 4 ± 1 °C. (**a**) Young's modulus and (**b**) hardness. Means with different letters (a, b, c and d) indicate significant differences ($p < 0.05$) during the shelf life for each single treatment.

From the 21st day of storage, the hardness decreased significantly ($p < 0.05$) (Figure 2b); this decrease was probably due to the initiation of the degradation processes, mainly related to autolytic phenomena and protein denaturation in muscle tissue, resulting in a progressive reduction in tissue hardness [19,58,59].

3.1.3. Water Holding Capacity (WHC)

The WHC is considered one of the most important parameters for preserving the fish quality and can influence the appearance and texture of fresh and processed fish products and, thus, the sensory quality of food [51,60,61].

Changes in the WHC detected in fresh and smoked meagre during their shelf life are shown in Figure 3.

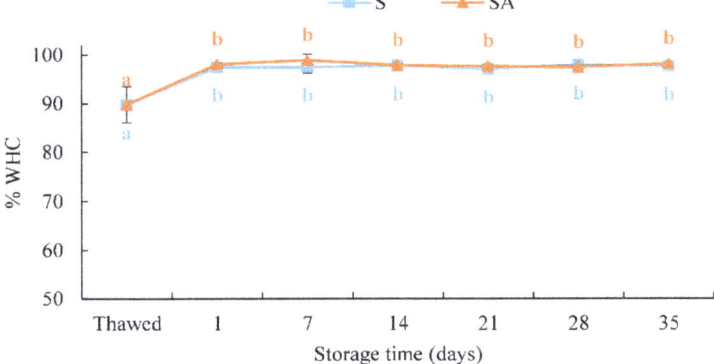

Figure 3. Results of the WHC% of cold-smoked meagre samples during storage at 4 ± 1 °C. (Means with different letters (a, b) indicate significant differences ($p < 0.05$) during the shelf life for each single treatment.

The smoking process resulted in a significant ($p < 0.05$) increase in the WHC in both treatments (Figure 3). This finding is in accordance with the results obtained in other cold-smoked species [19,22,54]. As observed by some authors [19,22,54], the increase in the

WHC in smoked products is the result of the increase in the salt content. Indeed, as it is shown in Table 2 after cold smoking, an increase in the percentage of ash in the muscle was measured as a consequence of the salt uptake in the fish muscles.

Table 2. The proximate composition (g/100 g) of thawed and cold-smoked meagre samples during storage at 4 ± 1 °C.

	Storage Time (Days)	S	SA
Total Lipids	Thawed	0.85 ± 0.06 [a]	0.88 ± 0.12 [a]
	1	1.72 ± 0.26 [b]	2.15 ± 0.23 [b]
	35	2.25 ± 0.26 [b]	2.51 ± 0.39 [b]
Moisture	Thawed	79.57 ± 1.74 [b]	80.57 ± 1.79 [b]
	1	74.96 ± 1.54 [a]	74.19 ± 1.05 [a]
	35	73.03 ± 1.17 [a]	73.08 ± 1.71 [a]
Protein	Thawed	18.06 ± 1.24	17.04 ± 1.18
	1	18.18 ± 0.10	18.02 ± 0.10
	35	18.84 ± 0.53	18.84 ± 0.81
Ash	Thawed	1.26 ± 0.07 [a]	1.23 ± 0.07 [a]
	1	5.13 ± 0.88 [b]	5.22 ± 0.68 [b]
	35	5.10 ± 0.15 [b]	4.67 ± 0.88 [b]

The means with different letters (a, b) in the same column are significantly different ($p < 0.05$).

The WHC remained stable throughout the storage time, with no significant differences between the two treatments.

3.1.4. Muscular pH

Changes in the pH values of cold-smoked meagre fillets during its shelf life are shown in Figure 4.

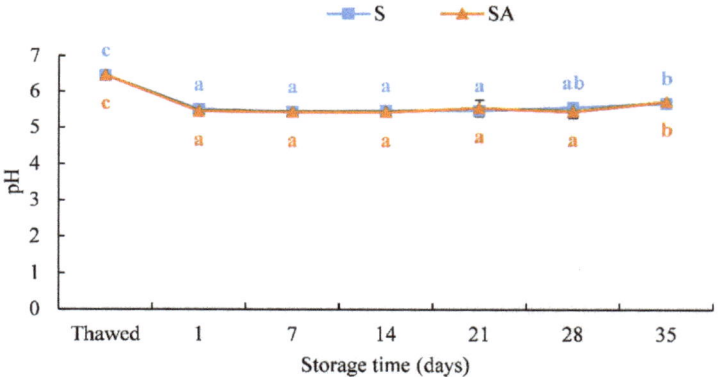

Figure 4. Results of the pH of cold-smoked meagre samples during storage at 4 ± 1 °C. (Means with different letters (a, b and c) indicate significant differences ($p < 0.05$) during the shelf life for each single treatment.

The smoking process significantly reduced the pH ($p < 0.05$). In fact, a significant difference was observed between the untreated and smoked samples ($p < 0.05$) (Figure 4).

The reduction in pH after smoking is related to the absorption of smoke acids; moisture loss and the reactions of phenols, polyphenols and carbonyl compounds with protein and amine groups [54,62]. In addition, the presence of salt, which causes an increase in the ionic strength of the solution within the cells, also contributes to the decrease in pH [54,63].

During the shelf life, it was observed that the pH remained constant in both treatments (Figure 4) and increased significantly on the last day of storage (day 35) with values of 5.69 ± 0.14 (S) and 5.73 ± 0.19 (SA).

The increase in pH can be attributed to the production of basic volatile components such as ammonia, trimethylamine and total volatile nitrogen by fish spoilage bacteria [22,64,65].

3.1.5. Water Activity Determination

The smoking process led to a decrease in the aw values detected in meagre fillets (Figure 5), which is also related to moisture reduction and an increase in the ash and mineral contents, as also observed in other species [22,54].

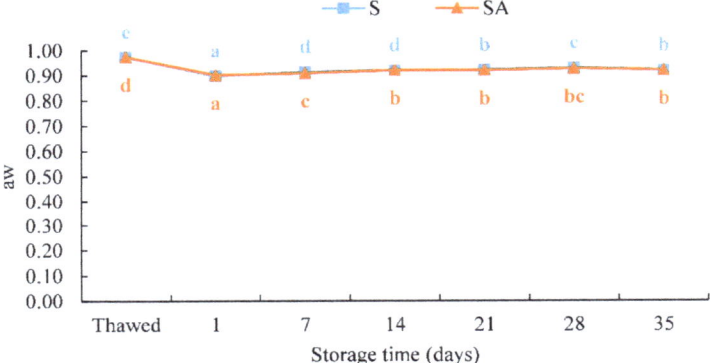

Figure 5. Results of the water activity (aw) of cold-smoked meagre samples during storage at 4 ± 1 °C. (Means with different letters (a, b, c, d and e) indicate significant differences ($p < 0.05$) during the shelf life for each single treatment.

Normally, the aw factor in fish is close to 1, a value that was also observed in thawed meagre fillets (Figure 5). Fish processing can determine the reduction in aw values, reaching even values of 0.8 and 0.7 after heavy salting and drying [20].

The decrease in aw, due to osmotic pressure, results in the lower activity of bacteria and enzymes [66].

3.2. Proximate Composition and Biochemical Parameters Related to the Shelf-Life

3.2.1. Proximate Composition

The proximate composition of untreated and smoked meagre fillets is shown in Table 2.

The lipid content showed that *A. regius* is a low-fat fish species [57] (Table 2).

In fact, farmed meagre has a much lower muscle fat than more commonly farmed Mediterranean fish species, including sea bream and European sea bass [68].

The relative contents of ash, water and lipids changed following smoking due to water loss. Salting and smoking determined a reduction in moisture and increase in the ash and mineral contents, as observed in previous studies on other fish species, such as European sea bass and dolphinfish [19,22,54].

Figure 6 shows the fatty acid profile of smoked meagre fillets.

The role played by the natural antioxidant in preserving the oxidation of the fatty acids was evident. In fact, S fillets (Figure 6) showed a significant reduction in polyunsaturated fatty acids ($p < 0.05$) at the end of smoking (35 days) due to a decrease in polyunsaturated fatty acids of the n-3 series (Tot n-3) ($p < 0.05$) and a simultaneous increase in monounsaturated fatty acids (Figure 6). In SA smoked sea bass fillets, Tot n-3 remained unchanged during the shelf life, as did the other classes of fatty acids.

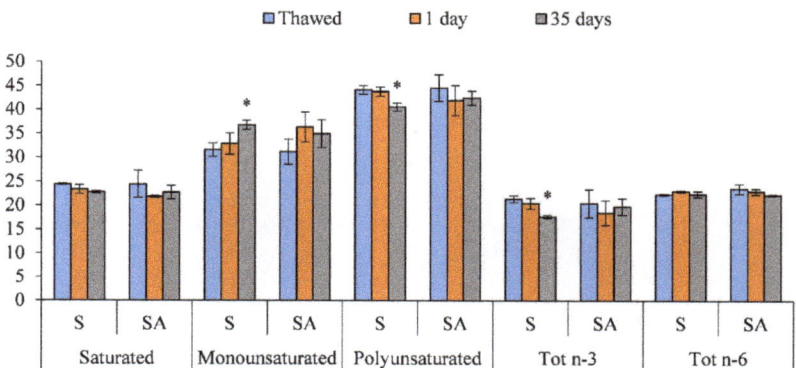

Figure 6. Fatty acid classes of thawed and cold-smoked meagre samples during storage at 4 ± 1 °C (* $p < 0.05$).

3.2.2. Biochemical Parameters Related to the Shelf-Life

The antioxidant treatment of fillets of different species has been shown to prevent lipid oxidation [19].

Malondialdehyde (MDA) is a secondary lipid oxidation product. As observed in other fish species, such as dolphinfish (*C. hippurus*), a lower lipid content, combined with a high polyunsaturated fatty acid content, could lead to a greater susceptibility to peroxidation than other fish species, such as sardines (*S. pilchardus*), which have a higher lipid content but a higher saturated fatty acid content, which are less susceptible to oxidation [27].

As shown in Figure 7, the SA treatment determined a marked reduction in the MDA content (mg MDA/kg) in smoked meagre fillets compared to the S treatment, which produced a higher MDA content after 7 days of refrigerated storage. The levels of the MDA results were statistically different between the treatments ($p < 0.05$) (Figure 7) from 7 days until the end of the trial.

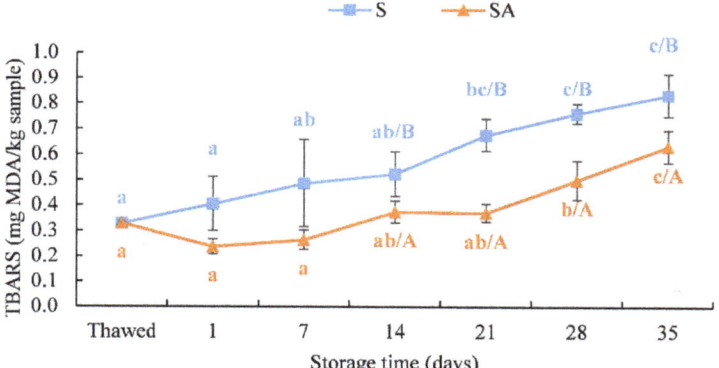

Figure 7. Results of the TBARS (determined as mg MDA/kg samples) of cold-smoked meagre samples during storage at 4 ± 1 °C. (Means with different letters (a, b and c) indicate significant differences ($p < 0.05$) during the shelf life for each single treatment. Means with different letters (A, B) indicate significant differences ($p < 0.05$) between the two treatments at the same storage time.

The low MDA content was detected in sous vide meagre (*A. regius*) fillets treated with natural antioxidants during cold storage, and the MDA values did not exceed the limit value during the storage period [69]. In that study, low values of MDA (0.52 ± 0.13-mg

MDA/kg) were detected in the raw fillets, and lower values of MDA were found during cold storage in fillets treated with antioxidants compared to the control, highlighting that the application of natural antioxidants in the meagre sous vide process could have had positive effects on the lipid peroxidation process [69].

Moreover, the results obtained in the present study in terms of the low MDA content detected in smoked fillets during storage agreed with those obtained in another farmed species, i.e., European sea bass (*D. labrax*), whose fillets were treated by cold smoking, highlighting the preservative effect of the process on this biochemical parameter related to the fish shelf life [22].

The results obtained in the present study also agreed with what was obtained in previous studies, in which the antioxidant treatment combined with smoking was effective in preventing lipid peroxidation [19,27]. Indeed, the combined treatments markedly reduced the MDA content in dolphinfish fillets up to 35 days of storage compared to cold smoking used alone as a preservative treatment [19]. Moreover, as in the case of sardines (*S. pilchardus*), a fatty fish species, the lipid oxidation results showed that the combined application of oregano extract and smoking significantly reduced the oxidation ($p < 0.05$) in sardines in comparison with the batch that was only smoked [27].

It has to be highlighted that, in both cases, cold-smoked fillets did not exceed the 4 to 5-mg MDA/kg value considered acceptable for smoked fish in the literature [65]. Moreover, vacuum packaging might be effective in preserving processed products from lipid oxidation, from a technology hurdle point of view, providing more evidence on the effective use of these methods of preservation [70,71].

The total volatile basic nitrogen (TVB-N) values detected in smoked meagre fillets during storage are shown in Figure 8. European legislation has established an upper limit for TVB-N, ranging from 25 to 35-mg TVB-N/100 g [42]. Nevertheless, considering that a TVB-N threshold for processed fish products has not been established so far, a limit equal to 35-mg TVB-N/100 g has been considered for smoked fish products [19,25]. In the present study, this threshold was not exceeded during the preservation period and up to 35 days of storage in both batches (SA and S) (Figure 7). Nevertheless, it has to be highlighted that, at T35, the level of TVB-N content in the SA batch was lower than that observed in the S batch, with their values significantly different ($p < 0.05$).

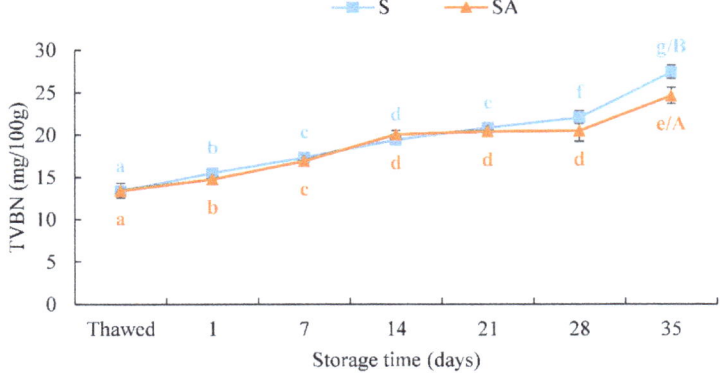

Figure 8. TVBN contents in cold-smoked meagre fillets during storage at 4 ± 1 °C. (Means with different letters (a, b, c, d, e, f and g) indicate significant differences ($p < 0.05$) during the shelf life for each single treatment; means with different letters (A, B) indicate significant differences ($p < 0.05$) between the two treatments at the same storage time.

The TVB-N may be affected by a reduction in both the spoilage bacteria and the activity of endogenous enzymes [72]. The findings obtained in the present study agreed with what was obtained in previous studies, which highlighted the effectiveness of the salting/smoking processes on maintaining the TVB-N content below the threshold for spoilage during refrigerated storage under vacuum or modified atmosphere packaging [19,22,73].

3.3. Microbiological Analyses

Microbiological analyses highlighted the bacteriostatic effect of cold smoking combined with the natural antioxidant. This effect is due to the synergistic action of the salt uptake during the brining phase and the polyphenols deposition during the smoking phase [19,25,74]. In thawed meagre fillets, the bacterial load was 4.40 ± 0.29 log (CFU/g) for mesophilic bacteria and 4.40 ± 0.89 log (CFU/g) for psychrophilic bacteria (Figure 9). During storage, the bacterial load remained constant in both treatments up to 21 and 28 days of storage for mesophilic bacteria and psychrophilic bacteria, respectively. At the end of storage, a significant increase in both mesophilic and psychrophilic bacteria was observed (Figure 9) in the two treatments, although the values did not exceed 10^7 CFU/g, according to the current standard [75–79].

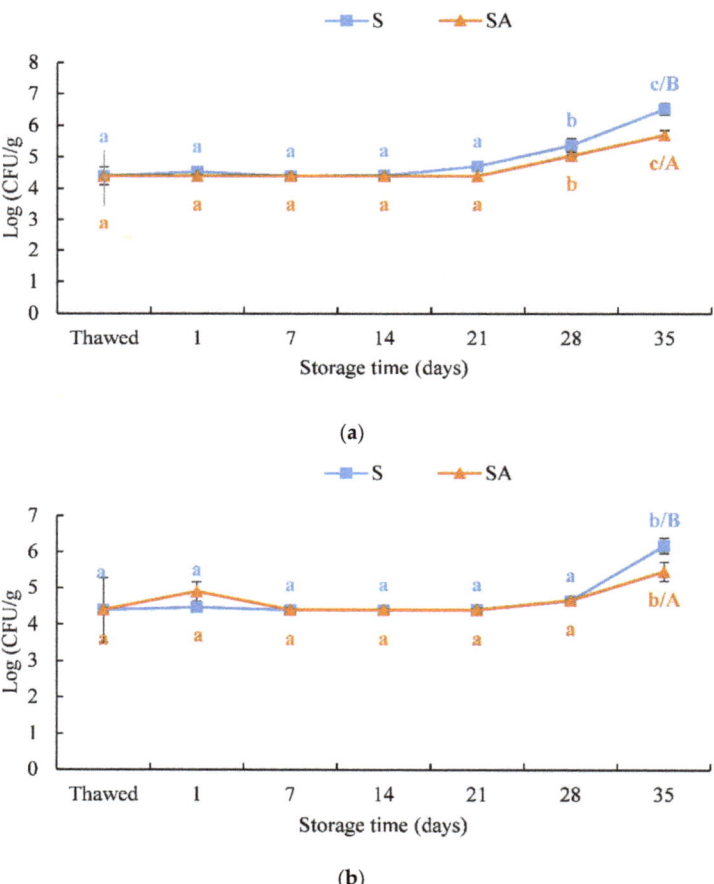

Figure 9. Microbiological evaluation during the shelf life of cold smoked meagre fillets stored at 4 ± 1 °C. (**a**) Mesophilic (37 °C) and (**b**) psychrophilic (6 °C). Means with different letters (a, b and c) indicate significant differences ($p < 0.05$) during the shelf life for each single treatment. Means with different letters (A and B) indicate significant differences ($p < 0.05$) between the two treatments at the same storage time.

3.4. Sensorial Analysis

The results from the sensory evaluation were strictly related to those of the physical–chemical and biochemical parameters considered to determine the fish quality and freshness during the shelf-life trials and give important findings about consumer perception and acceptability [19,25,80]. In the present study, both the S and SA batches did not show any differences for any of the sensory attributes evaluated, showing that the method of processing and the use of antioxidants from the halophyte *H. strobilaceum* effectively preserved the quality and shelf life of meagre [19]. The sensory quality evaluation showed that the smoked fillets of meagre were of good quality until 35 days of refrigerated storage under vacuum packaging (Table 3).

Table 3. Results of the sensory analysis of cold-smoked meagre samples during storage at 4 ± 1 °C.

	Days	S	SA
Appearance	1	9.00 ± 0.82 [bcd]	9.00 ± 0.00 [bcd]
	7	10.00 ± 0.00 [d]	9.25 ± 0.50 [d]
	14	8.25 ± 0.50 [abc]	8.75 ± 0.50 [bcd]
	21	8.00 ± 1.15 [abc]	7.75 ± 0.50 [ab]
	28	8.00 ± 0.89 [abc]	8.17 ± 0.41 [abc]
	35	7.17 ± 0.75 [a]	7.00 ± 0.89 [a]
Color	1	9.25 ± 0.96 [cd]	8.75 ± 0.50 [bc]
	7	10.00 ± 0.00 [d]	9.25 ± 0.50 [c]
	14	8.50 ± 0.58 [bc]	8.75 ± 0.50 [bc]
	21	8.50 ± 0.58 [bc]	8.25 ± 0.96 [b]
	28	8.17 ± 0.98 [bc]	8.50 ± 0.55 [b]
	35	7.33 ± 0.52 [ab]	7.00 ± 0.89 [a]
Odor	1	9.50 ± 0.58 [d]	8.00 ± 0.00 [b]
	7	9.50 ± 0.58 [d]	9.25 ± 0.50 [b]
	14	8.50 ± 0.58 [bcd]	9.00 ± 0.82 [b]
	21	8.50 ± 0.58 [bcd]	8.00 ± 0.82 [b]
	28	7.83 ± 0.75 [bc]	8.17 ± 0.41 [b]
	35	7.17 ± 0.75 [ab]	6.67 ± 1.03 [a]
Texture	1	9.75 ± 0.50 [c]	7.50 ± 0.58 [a]
	7	9.75 ± 0.50 [c]	9.00 ± 0.00 [bc]
	14	9.00 ± 0.00 [bc]	9.75 ± 0.50 [c]
	21	8.50 ± 0.58 [b]	8.25 ± 0.96 [b]
	28	9.00 ± 0.00 [bc]	9.00 ± 0.00 [bc]
	35	7.33 ± 0.52 [a]	7.00 ± 0.89 [a]
Taste	1	9.50 ± 0.58 [b]	8.25 ± 0.50 [bc]
	7	9.50 ± 0.58 [b]	9.25 ± 0.50 [c]
	14	8.50 ± 0.58 [ab]	9.00 ± 0.82 [c]
	21	8.25 ± 0.96 [ab]	7.25 ± 0.96 [ab]
	28	7.83 ± 0.75 [a]	8.17 ± 0.41 [bc]
	35	7.17 ± 0.75 [a]	6.67 ± 1.03 [a]

The means with different letters (a, b, c and d) in the same column are significantly different ($p < 0.05$).

As far as odor and taste were concerned, the sensory evaluation performed confirmed what was obtained from the biochemical analyses in terms of the TVB-N and MDA contents. In particular, even if significant differences were observed between the batches in terms of the MDA contents, the minimum value of 1.44-mg MDA/kg detectable by the panelists [81] was not exceeded in both batches up to the end of storage (Figure 6).

4. Conclusions

A. regius is today one of the new frontiers of aquaculture. Our work demonstrated the validity of processing this species through cold smoking with the use of natural antioxidants, which could therefore represent a means of obtaining an innovative yet safe product.

In fact, the combined treatment with cold smoking and antioxidant resulted in an overall improvement in the quality characteristics and shelf life of meagre fillets; the antioxidant treatment prevented the lipid peroxidation and degradation of the nitrogen components.

From a sensory and technological point of view, the presence of antioxidants of natural origin did not lead to detectable changes in the final product.

Author Contributions: Conceptualization, A.S. and C.M.M.; methodology, A.S. and C.M.M.; validation, A.S. and C.M.M.; formal analysis, R.A., L.L.B., M.R., M.M. and G.F.; resources, A.S.; data curation, A.S. and C.M.M.; writing—original draft preparation, R.A. and G.F.; writing—review and editing, C.M.M., S.S. and A.S.; supervision, C.M.M. and A.S. and funding acquisition, A.S. All authors have read and agreed to the published version of the manuscript.

Funding: This research was partially supported by the project INNOVAQUA—Innovazione tecnologica a supporto dell'incremento della produttività e della competitività dell'acquacoltura siciliana-PON02_00667. The author G.F. was funded by the PON RI 2014–2020 Cod DOT1320192 CUP B75B17000580001 PhD fellowship: "Innovative technologies for packaging, preservation, shelf-life extension and traceability of the aquaculture and fishery products".

Institutional Review Board Statement: Not applicable.

Informed Consent Statement: Not applicable.

Data Availability Statement: Not applicable.

Conflicts of Interest: The authors declare no conflict of interest.

References

1. FAO. *The State of World Fisheries and Aquaculture 2020*; FAO: Rome, Italy, 2020.
2. Bonanomi, S.; Colombelli, A.; Malvarosa, L.; Cozzolino, M.; Sala, A. Towards the introduction of sustainable fishery products: The bid of a major Italian retailer. *Sustainability* **2017**, *9*, 438. [CrossRef]
3. EUMOFA. The EU Fish Market. European Market Observatory for Fisheries and Aquaculture Products. 2020. Available online: www.eumofa.eu (accessed on 9 July 2021).
4. Parisi, G.; Terova, G.; Gasco, L.; Piccolo, G.; Roncarati, A.; Moretti, V.M.; Centoducati, G.; Gatta, P.P.; Pais, A. Current status and future perspectives of Italian finfish aquaculture. *Rev. Fish Biol. Fish.* **2014**, *24*, 15–73. [CrossRef]
5. Poli, B.M.; Parisi, G.; Zampacavallo, G.; Iurzan, F.; Mecatti, M.; Lupi, P.; Bonelli, A. Preliminary results on quality and quality changes in reared meagre (*Argyrosomus regius*): Body and fillet traits and freshness changes in refrigerated commercial-size fish. *Aquac. Int.* **2003**, *11*, 301–311. [CrossRef]
6. Arab-Tehrany, E.; Jacquot, M.; Gaiani, C.; Imran, M.; Desobry, S.; Linder, M. Beneficial effects and oxidative stability of omega-3 long-chain polyunsaturated fatty acids. *Trends Food Sci. Technol.* **2012**, *25*, 24–33. [CrossRef]
7. Mei, J.; Ma, X.; Xie, J. Review on natural preservatives for extending fish shelf life. *Foods* **2019**, *8*, 490. [CrossRef] [PubMed]
8. Sallam, K.I. Antimicrobial and antioxidant effects of sodium acetate, sodium lactate, and sodium citrate in refrigerated sliced salmon. *Food Control* **2007**, *18*, 566–575. [CrossRef] [PubMed]
9. Lalithapriya, U.; Mariajenita, P.; Renuka, V.; Sudharsan, K.; Karthikeyan, S.; Sivarajan, M.; Murugan, D.; Sukumar, M. Investigation of Natural Extracts and Sodium Bisulfite Impact on Thermal Signals and Physicochemical Compositions of Litopenaeus vannamei during Chilled Storage. *J. Aquat. Food Prod. Technol.* **2019**, *28*, 609–623. [CrossRef]
10. Liau, B.-C.; Shen, C.-T.; Liang, F.-P.; Hong, S.-E.; Hsu, S.-L.; Jong, T.-T.; Chang, C.-M.J. Supercritical fluids extraction and anti-solvent purification of carotenoids from microalgae and associated bioactivity. *J. Supercrit Fluids* **2010**, *55*, 169–175. [CrossRef]
11. Yoshie-Stark, Y.; Hsieh, Y. Distribution of flavonoids and related compounds from seaweeds in Japan. *J. Tokyo Univ. Fish.* **2003**, *89*, 1–6.
12. Herrero, M.; Cifuentes, A.; Ibañez, E. Sub- and supercritical fluid extraction of functional ingredients from different natural sources: Plants, food-by-products, algae and microalgae—A review. *Food Chem.* **2006**, *98*, 136–148. [CrossRef]
13. Anaëlle, T.; Serrano Leon, E.; Laurent, V.; Elena, I.; Mendiola, J.A.; Stéphane, C.; Nelly, K.; Stéphane, L.B.; Luc, M.; Valérie, S.-P. Green improved processes to extract bioactive phenolic compounds from brown macroalgae using Sargassum muticum as model. *Talanta* **2013**, *104*, 44–52. [CrossRef]
14. Burritt, D.J.; Larkindale, J.; Hurd, C.L. Antioxidant metabolism in the intertidal red seaweed Stictosiphonia arbuscula following desiccation. *Planta* **2002**, *215*, 829–838. [CrossRef]
15. Hassoun, A.; Emir Çoban, Ö. Essential oils for antimicrobial and antioxidant applications in fish and other seafood products. *Trends Food Sci. Technol.* **2017**, *68*, 26–36. [CrossRef]
16. Kharchoufi, S.; Licciardello, F.; Siracusa, L.; Muratore, G.; Hamdi, M.; Restuccia, C. Antimicrobial and antioxidant features of 'Gabsi' pomegranate peel extracts. *Ind. Crops Prod.* **2018**, *111*, 345–352. [CrossRef]

17. Messina, C.M.; Renda, G.; Laudicella, V.A.; Trepos, R.; Fauchon, M.; Hellio, C.; Santulli, A. From ecology to biotechnology, study of the defense strategies of algae and halophytes (from Trapani Saltworks, NW Sicily) with a focus on antioxidants and antimicrobial properties. *Int. J. Mol. Sci.* **2019**, *20*, 881. [CrossRef] [PubMed]
18. Messina, C.M.; Bono, G.; Renda, G.; La Barbera, L.; Santulli, A. Effect of natural antioxidants and modified atmosphere packaging in preventing lipid oxidation and increasing the shelf-life of common dolphinfish (*Coryphaena hippurus*) fillets. *LWT—Food Sci. Technol.* **2015**, *62*, 271–277. [CrossRef]
19. Messina, C.M.; Bono, G.; Arena, R.; Randazzo, M.; Morghese, M.; Manuguerra, S.; La Barbera, L.; Ozogul, F.; Sadok, S.; Santulli, A. The combined impact of cold smoking and natural antioxidants on quality and shelf life of dolphinfish (*Coryphaena hippurus*) fillets. *Food Sci. Nutr.* **2019**, *7*, 1239–1250. [CrossRef] [PubMed]
20. Sampels, S. The effects of processing technologies and preparation on the final quality of fish products. *Trends Food Sci. Technol.* **2015**, *44*, 131–146. [CrossRef]
21. Arvanitoyannis, I.S.; Kotsanopoulos, K. V Smoking of Fish and Seafood: History, Methods and Effects on Physical, Nutritional and Microbiological Properties. *Food Bioprocess Technol.* **2012**, *5*, 831–853. [CrossRef]
22. Messina, C.M.; Arena, R.; Ficano, G.; La Barbera, L.; Morghese, M.; Santulli, A. Combination of Freezing, Low Sodium Brine, and Cold Smoking on the Quality and Shelf-Life of Sea Bass (*Dicentrarchus labrax* L.) Fillets as a Strategy to Innovate the Market of Aquaculture Products. *Animals* **2021**, *11*, 185. [CrossRef]
23. Birkeland, S.; Bjerkeng, B. The quality of cold-smoked Atlantic salmon (*Salmo salar*) as affected by salting method, time and temperature. *Int. J. Food Sci. Technol.* **2005**, *40*, 963–976. [CrossRef]
24. Gómez-Estaca, J.; Giménez, B.; Gómez-Guillén, C.; Montero, P. Influence of frozen storage on aptitude of sardine and dolphinfish for cold-smoking process. *LWT—Food Sci. Technol.* **2010**, *43*, 1246–1252. [CrossRef]
25. Gómez-Guillén, M.C.; Gómez-Estaca, J.; Giménez, B.; Montero, P. Alternative fish species for cold-smoking process. *Int. J. Food Sci. Technol.* **2009**, *44*, 1525–1535. [CrossRef]
26. Cardinal, M.; Cornet, J.; Serot, T.; Baron, R. Effects of the smoking process on odour characteristics of smoked herring (*Clupea harengus*) and relationships with phenolic compound content. *Food Chem.* **2006**, *96*, 137–146. [CrossRef]
27. Gómez-Estaca, J.; Gómez-Guillén, M.C.; Montero, P. The effect of combined traditional and novel treatments on oxidative status of dolphinfish (*Coryphaena hippurus*) and sardine (*Sardina pilchardus*) muscle lipids. *Food Sci. Technol. Int.* **2013**, *20*, 431–440. [CrossRef]
28. Puke, S.; Galoburda, R. Factors affecting smoked fish quality: A review. In Proceedings of the Research for Rural Development, Jelgava, Latvia, 13–15 May 2020; Volume 35, pp. 132–139.
29. Sung, J.-H.; Park, S.-H.; Seo, D.-H.; Lee, J.-H.; Hong, S.-W.; Hong, S.-S. Antioxidative and Skin-Whitening Effect of an Aqueous Extract of Salicornia herbacea. *Biosci. Biotechnol. Biochem.* **2009**, *73*, 552–556. [CrossRef]
30. CIE. *Commission Internationale de l'Eclairage Colorimetry*, 3rd ed.; CIE: Vienna, Austria, 2004; Volume 15, pp. 1–82.
31. Commission Internationale de l'Eclairage. CIE Recommendations on Uniform Color Spaces, Color-Difference Equations, and Metric Color Terms. *Color Res. Appl.* **1977**, *2*, 5–6. [CrossRef]
32. Robertson, A.R. Historical development of CIE recommended color difference equations. *Color Res. Appl.* **1990**, *15*, 167–170. [CrossRef]
33. Messina, C.M.; Arena, R.; Manuguerra, S.; Renda, G.; Laudicella, V.A.; Ficano, G.; Fazio, G.; La Barbera, L.; Santulli, A. Farmed Gilthead Sea Bream (*Sparus aurata*) by-Products Valorization: Viscera Oil ω-3 Enrichment by Short-Path Distillation and In Vitro Bioactivity Evaluation. *Mar. Drugs* **2021**, *19*, 160. [CrossRef]
34. Orban, E.; Sinesio, F.; Paoletti, F. The Functional Properties of the Proteins, Texture and the Sensory Characteristics of Frozen Sea Bream Fillets (*Sparus aurata*) from different farming systems. *LWT—Food Sci. Technol.* **1997**, *30*, 214–217. [CrossRef]
35. Teixeira, B.; Fidalgo, L.; Mendes, R.; Costa, G.; Cordeiro, C.; Marques, A.; Saraiva, J.A.; Nunes, M.L. Effect of high pressure processing in the quality of sea bass (*Dicentrarchus labrax*) fillets: Pressurization rate, pressure level and holding time. *Innov. Food Sci. Emerg. Technol.* **2014**, *22*, 31–39. [CrossRef]
36. AOAC. *Official Methods of Analysis: Changes in Official Methods of Analysis Made at the Annual Meeting*; AOAC: Rockville, MD, USA, 1990; Volume 15, ISBN 9780935584752.
37. Association of Official Analytical Chemists (AOAC). *AOAC Official Method 981.10 Crude Protein in Meat Block Digestion Method*; AOAC: Rockville, MD, USA, 1992; Volume 65, p. 1339.
38. Folch, J.; Lees, M.; Stanley, G.H.S. A simple method for the isolation and purification of total lipids from animal tissues. *J. Biol. Chem.* **1957**, *226*, 497–509. [CrossRef]
39. Lepage, G.; Roy, C.C. Improved recovery of fatty acid through direct transesterification without prior extraction or purification. *J. Lipid Res.* **1984**, *25*, 1391–1396. [CrossRef]
40. Messina, C.M.; Renda, G.; La Barbera, L.; Santulli, A. By-products of farmed European sea bass (*Dicentrarchus labrax* L.) as a potential source of n-3 PUFA. *Biologia* **2013**, *68*, 288–293. [CrossRef]
41. Botsoglou, N.A.; Fletouris, D.J.; Papageorgiou, G.E.; Vassilopoulos, V.N.; Mantis, A.J.; Trakatellis, A.G. Rapid, Sensitive, and Specific Thiobarbituric Acid Method for Measuring Lipid Peroxidation in Animal Tissue, Food, and Feedstuff Samples. *J. Agric. Food Chem.* **1994**, *42*, 1931–1937. [CrossRef]

42. Commissione delle Comunità Europee. *95/149/CE: Decisione della Commissione, dell'8 marzo 1995, che fissa i Valori Limite di ABVT (Azoto Basico Volatile Totale) per Talune Categorie di Prodotti della Pesca e i Relativi Metodi D'analisi*; Commissione delle Comunità Europee: Bruxelles, Belgium, 1995.
43. Bilgin, Ş.; Değirmenci, A. Quality changes in reared, hot-smoked meagre (Argyrosomus regius asso, 1801) during chill storage at 4 ± 1 °C. *Food Sci. Technol.* **2019**, *39*, 507–514. [CrossRef]
44. Fuentes, A.; Fernandez-Segovia, I.; Barat, J.M.; Serra, J.A. Physicochemical characterization of some smoked and marinated fish products. *J. Food Process. Preserv.* **2010**, *34*, 83–103. [CrossRef]
45. Cheng, J.H.; Sun, D.W.; Zeng, X.A.; Liu, D. Recent Advances in Methods and Techniques for Freshness Quality Determination and Evaluation of Fish and Fish Fillets: A Review. *Crit. Rev. Food Sci. Nutr.* **2015**, *55*, 1012–1225. [CrossRef]
46. Choubert, G.; Blanc, J.-M.; Courvalin, C. Muscle carotenoid content and colour of farmed rainbow trout fed astaxanthin or canthaxanthin as affected by cooking and smoke-curing procedures. *Int. J. Food Sci. Technol.* **1992**, *27*, 277–284. [CrossRef]
47. Cardinal, M.; Knockaert, C.; Torrissen, O.; Sigurgisladottir, S.; Morkore, T.; Thomassen, M.; Vallet, J.L. Relation of smoking parameters to the yield, colour and sensory quality of smoked Atlantic salmon (*Salmo salar*). *Food Res. Int.* **2001**, *34*, 537–550. [CrossRef]
48. Birkeland, S.; Rørå, A.M.B.; Skåra, T.; Bjerkeng, B. Effects of cold smoking procedures and raw material characteristics on product yield and quality parameters of cold smoked Atlantic salmon (*Salmo salar* L.) fillets. *Food Res. Int.* **2004**, *37*, 273–286. [CrossRef]
49. Offer, G.; Knight, P.; Jeacocke, R.; Almond, R.; Cousins, T. The structural basis of the water-holding, appearance and toughness of meat and meat products. *Food Microstruct.* **1989**, *8*, 151–170.
50. Bekhit, A.E.D.A.; Morton, J.D.; Dawson, C.O.; Sedcole, R. Optical properties of raw and processed fish roes from six commercial New Zealand species. *J. Food Eng.* **2009**, *91*, 363–371. [CrossRef]
51. Chan, S.S.; Roth, B.; Jessen, F.; Løvdal, T.; Jakobsen, A.N.; Lerfall, J. A comparative study of Atlantic salmon chilled in refrigerated seawater versus on ice: From whole fish to cold-smoked fillets. *Sci. Rep.* **2020**, *10*, 1–12. [CrossRef]
52. Richards, M.P.; Hultin, H.O. Effect of pH on lipid oxidation using trout hemolysate as a catalyst: A possible role for deoxyhemoglobin. *J. Agric. Food Chem.* **2000**, *48*, 3141–3147. [CrossRef] [PubMed]
53. Fuentes, A.; Fernández-Segovia, I.; Serra, J.A.; Barat, J.M. Development of a smoked sea bass product with partial sodium replacement. *LWT—Food Sci. Technol.* **2010**, *43*, 1426–1433. [CrossRef]
54. Fuentes, A.; Fernandez-Segovia, I.; Serra, J.A.; Barat, J.M. Effect of partial sodium replacement on physicochemical parameters of smoked sea bass during storage. *Food Sci. Technol. Int.* **2012**, *18*, 207–217. [CrossRef]
55. Regost, C.; Jakobsen, J.V.; Rørå, A.M.B. Flesh quality of raw and smoked fillets of Atlantic salmon as influenced by dietary oil sources and frozen storage. *Food Res. Int.* **2004**, *37*, 259–271. [CrossRef]
56. Gómez-Guillén, M.C.; Montero, P.; Hurtado, O.; Borderías, A.J. Biological characteristics affect the quality of farmed atlantic salmon and smoked muscle. *J. Food Sci.* **2000**, *65*, 53–60. [CrossRef]
57. Sigurgisladottir, S.; Sigurdardottir, M.S.; Torrissen, O.; Vallet, J.L.; Hafsteinsson, H. Effects of different salting and smoking processes on the microstructure, the texture and yield of Atlantic salmon (*Salmo salar*) fillets. *Food Res. Int.* **2000**, *33*, 847–855. [CrossRef]
58. Hernández, M.D.; Martínez, F.J.; García García, B. Sensory evaluation of farmed sharpsnout seabream (*Diplodus puntazzo*). *Aquac. Int.* **2001**, *9*, 519–529. [CrossRef]
59. Liu, S.; Fan, W.; Zhong, S.; Ma, C.; Li, P.; Zhou, K.; Peng, Z.; Zhu, M. Quality evaluation of tray-packed tilapia fillets stored at 0 °C based on sensory, microbiological, biochemical and physical attributes. *Afr. J. Biotechnol.* **2010**, *9*, 692–701. [CrossRef]
60. Huff-lonergan, E.; Lonergan, S.M. MEAT Mechanisms of water-holding capacity of meat: The role of postmortem biochemical and structural changes. *Meat Sci.* **2005**, *71*, 194–204. [CrossRef]
61. Chan, S.S.; Roth, B.; Skare, M.; Hernar, M.; Jessen, F.; Løvdal, T.; Jakobsen, A.N.; Lerfall, J. Effect of chilling technologies on water holding properties and other quality parameters throughout the whole value chain: From whole fish to cold-smoked fillets of Atlantic salmon (*Salmo salar*). *Aquaculture* **2020**, *526*, 735381. [CrossRef]
62. Hassan, I.M. Processing of smoked common carp fish and its relation to some chemical, physical and organoleptic properties. *Food Chem.* **1988**, *27*, 95–106. [CrossRef]
63. Leroi, F.; Joffraud, J.J. Salt and smoke simultaneously affect chemical and sensory quality of cold-smoked salmon during 5 degrees C storage predicted using factorial design. *J. Food Prot.* **2000**, *63*, 1222–1227. [CrossRef]
64. Ruiz-Capillas, C.; Moral, A. Sensory and biochemical aspects of quality of whole bigeye tuna (*Thunnus obesus*) during bulk storage in controlled atmospheres. *Food Chem.* **2005**. [CrossRef]
65. Osheba, A.S. Technological attempts for production of low sodium smoked herring fish (Renga). *Adv. J. Food Sci. Technol.* **2013**, *5*, 695–706. [CrossRef]
66. Oliveira, H.; Pedro, S.; Nunes, M.L.; Costa, R.; Vaz-Pires, P. Processing of Salted Cod (*Gadus* spp.): A Review. *Compr. Rev. Food Sci. Food Saf.* **2012**, *11*, 546–564. [CrossRef]
67. Tang, S.; Kerry, J.P.; Sheehan, D.; Buckley, D.J.; Morrissey, P.A. Antioxidative effect of added tea catechins on susceptibility of cooked red meat, poultry and fish patties to lipid oxidation. *Food Res. Int.* **2001**, *34*, 651–657. [CrossRef]
68. Grigorakis, K.; Fountoulaki, E.; Vasilaki, A.; Mittakos, I.; Nathanailides, C. Lipid quality and filleting yield of reared meagre (*Argyrosomus regius*). *Int. J. Food Sci. Technol.* **2011**, *46*, 711–716. [CrossRef]

69. Bozova, B.; İzci, L. Acta Aquatica Turcica Effects of Plant Extracts on the Quality of Sous Vide Meagre (*Argyrosomus regius*) Fillets. *Acta Aquatica Turcica* **2021**, *17*, 255–266.
70. Hussain, M.A.; Sumon, T.A.; Mazumder, S.K.; Ali, M.M.; Jang, W.J.; Abualreesh, M.H.; Sharifuzzaman, S.M.; Brown, C.L.; Lee, H.T.; Lee, E.W.; et al. Essential oils and chitosan as alternatives to chemical preservatives for fish and fisheries products: A review. *Food Control* **2021**, *129*, 108244. [CrossRef]
71. Leistner, L.; Gorris, L.G.M. Food preservation by hurdle technology. *Trends Food Sci. Technol.* **1995**, *6*, 41–46. [CrossRef]
72. Özyurt, G.; Kuley, E.; Özkütük, S.; Özogul, F. Sensory, microbiological and chemical assessment of the freshness of red mullet (*Mullus barbatus*) and goldband goatfish (*Upeneus moluccensis*) during storage in ice. *Food Chem.* **2009**, *114*, 505–510. [CrossRef]
73. Fuentes, A.; Fernández-Segovia, I.; Barat, J.M.; Serra, J.A. Influence of sodium replacement and packaging on quality and shelf life of smoked sea bass (*Dicentrarchus labrax* L.). *LWT—Food Sci. Technol.* **2011**, *44*, 917–923. [CrossRef]
74. Oueslati, S.; Ksouri, R.; Falleh, H.; Pichette, A.; Abdelly, C.; Legault, J. Phenolic content, antioxidant, anti-inflammatory and anticancer activities of the edible halophyte Suaeda fruticosa Forssk. *Food Chem.* **2012**, *132*, 943–947. [CrossRef]
75. AFSSA. *AVIS de l'Agence Française de Sécurité Sanitaire des Aliments Concernant les Références Applicables aux Denrées Alimentaires en tant que Critères Indicateurs D'hygiène des Procédés*; AFSSA: Maisons-Alfort, France, 2008; pp. 1–21.
76. FCD. *Critères Microbiologiques Applicables à Partir de 2010 aux Marques de Distributeurs, Marques Premiers prix et Matières Premières dans leur Conditionnement Initial Industriel*; FCD: Paris, France, 2009; pp. 1–57.
77. Healt Protection Agency. *Guidelines for Assessing the Microbiological Safety of Ready-to-Eat Foods Placed on the Market*; Healt Protection Agency: London, UK, 2009.
78. Decreto Ministeriale del 18/09/2002. *Modalità di Informazione sullo Stato di Qualità delle Acque, del Decreto Legislativo 11 Maggio 1999, n. 152. Gazz. Uff. Suppl. Ordin. n° 245 del 18/10/2002*; Ministro dell'Ambiente: Rome, Italy, 2002; p. 31030.
79. D.S. Reg. Piemonte Allegato B Linee Guida per l'analisi del Rischio. Progett. Reg. "Analisi del Rischio Microbiol. Legato al Consum. di Aliment. Final. alla Riduzione dei Costi Anal. Approv. con Determ. della Dir. Sanità della Reg. Piemonte n.780 del 18 Ottobre 2011. 2013, pp. 1–41. Available online: http://www.regione.piemonte.it/governo/bollettino/abbonati/2011/51/attach/dddb200000780_830.pdf (accessed on 1 September 2021).
80. Messina, C.M.; Bono, G.; Arena, R.; Randazzo, M.; Manuguerra, S.; Santulli, A. Polyphenols from halophytes and modified atmosphere packaging improve sensorial and biochemical markers of quality of common dolphinfish (*Coryphaena hippurus*) fillets. *Food Sci. Nutr.* **2016**, *4*, 723–732. [CrossRef]
81. Ruiz-Capillas, C.; Moral, A. Residual effect of CO_2 on hake (*Merluccius merluccius* L.) stored in modified and controlled atmospheres. *Eur. Food Res. Technol.* **2001**, *212*, 413–420. [CrossRef]

Article

Biostimulation as a Means for Optimizing Fruit Phytochemical Content and Functional Quality of Tomato Landraces of the San Marzano Area

Youssef Rouphael [1], Giandomenico Corrado [1,*], Giuseppe Colla [2], Stefania De Pascale [1], Emilia Dell'Aversana [3], Luisa Ida D'Amelia [3], Giovanna Marta Fusco [3] and Petronia Carillo [3]

[1] Department of Agricultural Sciences, University of Naples Federico II, Via Università 100, 80055 Portici, Italy; youssef.rouphael@unina.it (Y.R.); depascal@unina.it (S.D.P.)
[2] Department of Agriculture and Forest Sciences, University of Tuscia, Via San Camillo de Lellis, 01100 Viterbo, Italy; giucolla@unitus.it
[3] Department of Environmental, Biological and Pharmaceutical Sciences and Technologies, University of Campania "Luigi Vanvitelli", Via Vivaldi 43, 81100 Caserta, Italy; emilia.dellaversana@unicampania.it (E.D.); luisa.damelia@unicampania.it (L.I.D.); giovannamarta.fusco@unicampania.it (G.M.F.); petronia.carillo@unicampania.it (P.C.)
* Correspondence: giandomenico.corrado@unina.it; Tel.: +39-0812-539-294

Abstract: The effect of plant biostimulation on fruits of traditional tomato germplasm is largely unknown. We examined how a tropical plant-derived biostimulant impacts the nutritional, functional, and compositional characteristics of tomato fruits from four landraces, collected in the San Marzano (SM) tomato Protected Designation of Origin (PDO) region, by profiling primary and secondary metabolites. Biostimulation was not able to completely reshuffle the morpho-physiological and nutritional profile of the four landraces. Their distinct phytochemical profile indicated a genotype-specific tuning of the analyzed traits, which also included an improved yield and fruit quality. Biostimulation of SM1 and SM3 increased photosynthetic accumulation of carbohydrate reserves, improved mineral nutrient use efficiency and consequently, yield (+21% and 34%, respectively). Moreover, biostimulation augmented the nutraceutical properties of the SM2 landrace. Interestingly, the plant-derived product increased in all genotypes lycopene, but not polyphenol accumulation in fruits. Our results show the potential of biostimulatory applications towards optimizing the fruit quality of the acclaimed SM landraces, which is suitable to satisfy both the rising consumer demand for premium traditional tomatoes and the technological needs of the food industry.

Keywords: *Solanum lycopersicum* L.; genetic variability; quality; food composition; biostimulant; plant tropical extract

1. Introduction

In the last few decades, the consumer appreciation of the importance of food quality has been steadily increasing, regardless of the limitations related to the public perception of a multi-dimensional attribute [1]. Recently, a growing awareness of the environmental impact of the food production (in terms, for instance, of pollution, greenhouse gas emission, soil depletion, biodiversity loss, and chemical pesticides) has led to the definition of sustainable food quality [2]. A more sustainable food production system implies the use of resources at a rate that can be tolerated and ultimately, fully replenished by our environment. Sustainable food quality should consequently cover various issues, including safety, affordability, and nutritional and functional values, while controlling the use of chemical fertilizers, herbicides, and pesticides by leveraging natural plant defenses and biodiversity. The transition to a more sustainable food quality requires the promotion of agri-food systems and consumer behaviors that do not only emphasize aesthetic, nutritional, and functional attributes [3]. It is also necessary to put emphasis on safeguarding plant genetic resources to reverse agro-biodiversity loss, and reducing dependence on synthetic chemicals to limit the environmental impact of agriculture [4]. Under these perspectives, crop landraces are gaining popularity because of their perceived distinctive features and gastronomic value, as well as their amenability to more sustainable production systems (e.g., organic farming and low-input agriculture) and short food supply chains [5].

Tomato (*Solanum lycopersicum* L.) is one of the most widely grown vegetables, globally consumed fresh and in a variety of processed products. This species was domesticated in the Americas and, after the Columbian exchange, Italy and Spain have been recognized as secondary centers of diversification [6]. In these countries, several tomatoes with fruit shapes and colors different from the domesticated forms have been documented since its introduction [6]. Locally adapted tomato landraces can have interesting traits such as resistance to stress and high-quality fruits [7–9]. Contemporary varieties often out-yield landraces. Consequently, strengthening the use of landraces towards sustainable fruit quality necessitates strategies and tools to guarantee provenance and authenticity (e.g., for premium brands) [10–12], as well as non-regulatory initiatives in order to overcome technical barriers and to highlight commercially valuable features [5]. A sustainable strategy for increasing yield in landraces could be the use of plant biostimulants (PB). PBs are substances or microorganisms, not classified as fertilizer or pesticide, that can increase resource efficiency, growth, yield, and abiotic stress resilience and tolerance when applied to plants [13]. PBs mainly act on plants by inducing direct and indirect multiple physiological effects, which are linked, to name a few, to an increased mobility and solubility of mineral nutrients in soil, changes in root system architecture, improved water use efficiency, and enhanced ion uptake, mobilization and use [14]. With respect to the composition and properties of plant food, PBs can increase the synthesis and accumulation of primary and secondary metabolites, including important categories of antioxidants, such as carotenoids, polyphenols, and ascorbic acid, thus ultimately improving the nutritional and nutraceutical quality of the edible products [15–17]. Moreover, biostimulants can be also used for biofortification, for instance, improving the mineral content of leaves and fruits as well as their functional profile [18–20]. Biostimulants comprise various categories such as organic substances (e.g., humic acids, protein hydrolysates, chitosan, vitamins), inorganic compounds (e.g., cobalt, silica, selenium), and plant growth promoting microorganisms and their extracts (e.g., fungi, algae, bacteria) [13]. Nonetheless, it should not be overlooked that innovative products for agriculture, especially those suitable also for organic farming, should have the added benefit of making use of raw material that is disposed as organic waste of plant origin (plant bio-waste), to foster sustainable agricultural growth with minimal problems of biological and chemical safety [4,21].

In this work we tested the ability of a plant-derived biostimulant to enhance the nutritional, functional, and compositional characteristics of tomato fruits. While previous studies focused on modern, often hybrid, high-yield varieties bred for intensive agriculture [17], our aim was to understand the effect of biostimulation on a traditional germplasm that is culturally and gastronomically linked to a specified region. We focused on a biostimulant extracted from the biowaste of a tropical plant, which can be used in organic farming. We employed four distinct, indeterminate landraces whose fruit shapes typify the range of variability present in the whole peeled tomatoes grown in the area designated to produce the Protected Designation of Origins (PDO) San Marzano berries [22,23]. The interaction between the plant-based biostimulant and the different landraces over the fruit quality was assessed considering yield, as well as the mineral content, starch, soluble sugars, amino acids, proteins, lycopene, anthocyanins, and polyphenols of the fruits.

2. Materials and Methods

2.1. Plant Material, Growth Conditions and Experimental Design

This work was carried out on four tomato (*Solanum lycopersicum* L.) landraces collected in the area designated to produce the "Pomodoro San Marzano dell'Agro Sarnese-nocerino DOP", the Protected Denomination of Origin (PDO) EU label scheme for the original San

Marzano tomato [23]. These landraces were therefore named SM1, SM2, SM3, and SM4. The experiment was carried out in the 2017 summer growing season, in a greenhouse covered with a 0.25 mm thick ethyl vinyl acetate sheet at the experimental pilot farm "Torre Lama" (Bellizzi, SA) of the Department of Agricultural Sciences. The main physical and chemical soil characteristics at the experimental farm were clay loam texture with the following proportion of sand, silt, and clay: 47%, 25%, and 28%. The soil electrical conductivity (EC) was 0.15 dS m^{-1} and the pH 7.8. Total nitrogen was 0.11%, and organic matter: 1.23% (w/w). The Olsen phosphorus and exchangeable potassium were 85 and 889 mg kg^{-1}, respectively. The tomato seedlings were transplanted on 2 May at the three-true-leaf stage. Plants grew under natural light conditions. The mean air temperature and relative humidity inside the greenhouse were 27.6 °C and 59%, respectively. Fertilizer was applied though drip irrigation system consisting of irrigation tubes placed 5 cm apart from the tomato plants, with holes spaced 0.3 m from each other, and an irrigation flow of 2.4 L·h^{-1}. The NS delivered through fertigation was made (in mM) of N-NO$_3^-$: 12.0; S: 1.5; P: 1.0; K: 6.0; Ca: 3.5; Mg: 2.3; N-NH$_4^+$: 1.0. Micronutrients in the NS were (in μM): Fe: 20; Mn: 9; Cu: 0.3; Zn: 1.6; B: 20; Mo: 0.3. The pH and electric conductivity of the NS were 6.4 and 1.9 dS·m^{-1} at 25 °C, respectively. Fertigation was performed once per day. Landraces were treated by foliar application with a NS containing the commercial biostimulant Auxym® (Italpollina, Rivoli Veronese, Italy), using as control treatment a no-biostimulant NS. The plant extract (PE) Auxym® is produced through water extraction and fermentation of tropical plant biomass and its composition is presented elsewhere [24]. Plants were sprayed uniformly (four treatments) with a NS containing 2 mL·L^{-1} of Auxym® using (PE treatment) or the NS (control treatment) using a 25-L tank weed sprayer, starting from the 35th day after the transplant (DAT). The experiment was carried out through a completely randomized design, namely four landraces (L), two biostimulant treatments (B), three replicates (R), resulting in 24 experimental units (4L × 2B × 3R).

2.2. Yield and Morphological Analysis

Considering the indeterminate growth pattern of the tomato landraces, fruit were continuously harvested starting 60 DAT (July 1) and continued until the end of the experiment (August 1; 90 DAT). During the harvest period, the marketable yield consisting of fully ripened fruits (mature red stage) was calculated on five plants located at the central part of each experimental unit. The fruits of the third truss were analyzed for quality parameters. Ten fresh fruits of each experimental unit were used for the biometric measurements (i.e., shape index, juice pH, and mineral content). The shape index of fruits was determined as a ratio of the maximum height length to maximum width, relative to the longitudinal section. The remaining subsamples were immediately frozen in liquid nitrogen and stored at −80 °C for further biochemical analyses.

2.3. Fruit Juice pH, Dry Matter Percentage and Ion Exchange Chromatography

Immediately after harvest, ten fresh tomato fruits of each experimental unit were homogenized in a blender (2 L; Waring HGB140, CA, USA) for one minute at low speed. The slurry was filtered through a two-layer cheesecloth and the juice pH was read with a digital pH meter (HI-9023; Hanna Instruments, Padua, Italy). The fruits' dry matter percentage was also determined as a percentage of fresh mass following fruit desiccation to constant weight in a forced-air oven at 75 °C for 72 h, and weighed (Denver Instruments, Denver, CO, USA). Dried fruit tissues were ground in a Wiley Mill to pass through an 841 μm mesh and used for mineral analysis. For fruit mineral profile and citrate analysis, 250 mg of the dried material were suspended in 50 mL of ultrapure water (Milli-Q, Merck Millipore, Darmstadt, Germany), subjected to three freeze-thaw cycles in liquid nitrogen, centrifuged for 10 min at 6000 rpm (R-10 M, Remi Elektrotechnik Limited, India) and filtered through a 0.20 μm filter Whatman paper (Whatman International Ltd., Maidstone, UK). The clear supernatant was assayed by ion-exchange chromatography (ICS-3000, Dionex, Sunnyvale, CA, USA) as described [24]. Results of mineral composition were expressed in g kg^{-1} dw, except for nitrate that was converted to mg kg^{-1} fw based on each sample's dw.

2.4. Starch and Soluble Carbohydrates Analysis

Starch and soluble sugars were determined as described by Carillo et al. [25] with some modifications. Fruits were frozen, finely ground, and 20 mg of powdered tissue were suspended in 250 µL of ethanol (98% v/v 5 mM Hepes/KOH; pH 7.0), incubated in a hot water bath (80 °C) for 20 min, and centrifuged at 14,000× g for 10 min at 4 °C. The clear supernatant was stored on ice. The pellet was submitted to two further extractions, first with 250 µL of 80% ethanol (v/v, 5 mM Hepes/KOH; pH 7.0), and then with 150 µL of 50% ethanol (v/v, 5 mM Hepes/KOH; pH 7.0), incubated and centrifuged as above. The supernatants of the three consecutive extractions were pooled and stored at -20 °C. The remaining pellet of the three ethanol extractions was used for starch determination. After the addition of 250 µL of 0.1 M KOH, samples were left at 90 °C for 2 h and then placed on ice. After cooling, the sample pH was brought to 4.5 by adding 75 µL of 1 M glacial acetic acid. An aliquot (100 µL) of acidified samples was added to 100 µL of 50 mM sodium acetate (pH 4.8) with 0.2 units α-amylase and 2 units amyloglucosidase and incubated at 37 °C for 18 h for starch hydrolysis. After centrifugation (14,000 rpm for 10 min at 4 °C), the soluble carbohydrates (fructose, glucose, sucrose) were analysed as previously detailed [26].

2.5. Antioxidant Metabolites Analysis

For anthocyanin quantification, frozen samples were finely powdered, and 40 mg were suspended in 200 µL of 40% (v/v) ethyl alcohol, thoroughly mixed, and incubated on ice for 20 min. After cold centrifugation (14,000 rpm, 10 min), the pellet was immediately extracted again using the same procedure. The two supernatants were then joined. Duplicate aliquots (150 µL) were dispensed in a microplate. Then, 75 µL of 25 mM KCl (pH 1.0) or 75 µL of 400 mM sodium acetate (pH 4.5) were added. A 150 µL of a 40% (w/v) ethanol solution was used as no-sample blank. The absorbance was read at 520 and 700 nm using a microplate-reading spectrophotometer (FLX-Xenius, SAFAS, Monaco, Germany). Quantification was performed as already reported [27]. Total anthocyanin content was expressed as mg cyanidin-3-glucoside equivalents per gram of fresh weight (mg C3G eq·g^{-1} FW).

Lycopene was evaluated using a previously published procedure with some modifications [28]. Powdered samples (20 mg) were suspended in 380 µL of solution of hexane, acetone, and methanol in a volume ratio of 2:1:1, containing also 0.05% (w/v) butylated hydroxytoluene. A no-plant extract was used as blank. The suspension was mixed on an orbital shaker for 30 min, centrifuged at 4 °C for 10 min at 14,000 rpm, and 100 µL of the orange organic phase were mixed with 1.4 mL hexane in a clean tube. The absorbance at 472 nm was measured as described above. Lycopene concentration was extrapolated with a calibration curve built with pure lycopene within the standard range (i.e., linear portion of the calibration curve) of 0.5–3 mg·L^{-1}. Lycopene quantity was converted in mg·g^{-1} FW.

Polyphenols were determined as reported with some modifications [29]. Powdered samples (20 mg) were suspended in 800 µL of 60% methanol. The suspension was shaken at 800 rpm in a vortex mixer for 15 min and then centrifuged for 5 min at 8000× g. Aliquots (100 µL) of the clear supernatant or the same volume of 60% methanol (blank) were added to 50 µL of the Folin–Ciocalteu reagent diluted with distilled water (1:1 v/v). Samples were shaken for 6 min at 800 rpm, and then 650 µL of 3% (w/v) sodium carbonate were added. After an incubation of one hour and a half at room temperature, sample absorbance was spectrophotometrically read at 760 nm as described above. Polyphenols were quantified using a standard curve of gallic acid (GAE) in the 25–125 mg·L^{-1} range and expressed as mg GAE equivalents per gram of fresh weight (mg GAE·g^{-1} FW).

2.6. Quantification of Soluble Proteins and Free Amino Acids

The extraction of the soluble proteins was carried out starting from 20 mg of frozen tissue incubated at 4 °C for 24 h in 500 µL of Tris-HCl 200 mM (pH 7.5) containing 500 mM $MgCl_2$. Samples were then centrifuged at $16,000 \times g$ for 10 min at 4 °C. For each biological replicate, triplicate aliquots (20 µL) of the clear supernatant, each with 180 µL of paper-filtered diluted (1:5 v/v) Protein Assay Dye Reagent Concentrate (Bio-Rad, Milan, Italy), were transferred to a 96–Well Flat–Bottom microtiter plate and thoroughly mixed. Protein standards (20 µL of 15, 37.5, 75, 112.5 and 150 mg·L^{-1}, corresponding to 0.3, 0.75, 1.5, 2.25 and 3 µg of lyophilised BSA) were diluted in 200 mM Tris-HCl containing 500 mM $MgCl_2$ as the samples. After incubation for at least 5 min at room temperature, absorbance was measured at 595 nm using a multi-detection microplate reader (Synergy HT, Biotek, Germany). Quantities were estimated using a blank corrected standard curve built with BSA and were expressed in mg·g^{-1} FW.

Amino acids were extracted essentially as described [30]. Samples (40 mg of powdered fruit tissue) were mixed with 1 mL of a 40% ethanol solution and left overnight at 4 °C. After cold-centrifuged at 14,000 rpm for 10 min. Primary AAs were analysed with a Nexera X2 Ultra High-Performance Liquid Chromatograph (Shimadzu, Milan, Italy), after an automated in needle three min pre-column derivatization of 20 µL of clear supernatant with 40 µL of derivatization solution [25]. Derivatized samples were injected onto a ZORBAX Eclipse Plus column (C18, 95 Å, 5 µm, 4.6 × 250 mm; Agilent Technologies, Milan, Italy) and eluted with a discontinuous gradient at 25 °C, with a 1 mL·min^{-1} flow rate. The detection of the amino acids (OPA-derivatized) was carried out using an excitation wavelength of 330 nm and reading emission at 450 nm. Peaks were assessed and quantified by comparing their relative retention time (RTT) and relative peak area (RPA) with that of injected reference standards [25]. Proline was quantified using the extract employed for the amino acids determination, employing an acid ninhydrin method according to a procedure previously described [31]. The amino acids were expressed as µmol·g^{-1} FW.

2.7. Net Economic Benefits: Partial Budget Analysis

A partial budget analysis was carried out essentially as described [18,32]. Briefly, a tomato selling price of 500 € t^{-1} at shipping point was used to calculate the added gross return of biostimulant-treated tomato production in comparison with untreated-tomato production. Moreover, the added variable costs (biostimulant product, foliar spraying, and fruit harvest of additional yield) were determined considering (i) a biostimulant selling price of 24 € L^{-1}, (ii) a cost of single foliar spraying of 100 € ha^{-1}, and (iii) hand-harvesting cost of 200 € t^{-1} [18]. The added net return was calculated as the difference between added gross return and added variable costs.

2.8. Statistical Analysis

A two-way Analysis of Variance (ANOVA) was carried out to examine the influence of the biostimulant treatment (B), the landrace (L), and their interaction (L × B). All data are presented as the mean ± Standard Error (SE). In the absence of significant L × B, mean separation was performed by Duncan's Multiple Range Test ($p < 0.05$) for L and by Student's t-test ($p < 0.05$) for B. For variables that were subject to significant L × B interaction, one-way ANOVA was performed according to Duncan's multiple range test ($p < 0.05$). Calculations were performed using the SPSS 20 software (IBM, Akron, NY, USA). The Principal Component Analysis (PCA) was performed using the Minitab 18 statistical software (Minitab LLC, State College, PA, USA) [33].

3. Results and Discussion

3.1. Effects on Yield, Ecomomic Profitability and on Chemical and Physical Fruit Characteristics of the Genotype, Biostimulation and Their Interaction

Seeds of tomato landraces were collected from small farms in the area designated to produce the San Marzano PDO, and then multiplied and selected for uniformity at the Department of Agricultural Sciences, University of Naples Federico II. The four landraces under investigation were chosen as each had a distinctive fruit phenotype (Supplementary Table S1), thus representing a suitable panel of fruit diversity of the local landraces. The effect of the biostimulant over the tomato yield and its chemical and physical characteristics is reported in Table 1. Significant differences due to the factor "Landrace" (L) were present for the total yield and all the other parameters, except for the dry matter of the fruit, a parameter important mainly to produce tomato concentrated paste and puree [34]. Although trait variation was limited, the observed differences further indicated that, in addition to different fruit shapes, the four landraces have distinctive features. For the measured parameters, a significant interaction between the genotype and plant extract application (B) was not observed (Table 1). The main effect of the biostimulant treatment was more complex. Overall, the plant extract had a small positive effect on yield, but factors' interaction had a main role, because yield increase was different among landraces. Specifically, the PE treatment of the two landraces with the higher yield, SM1 and SM3, increased this parameter of 21% and 35%, respectively. Overall, the added marketable yield (averaged across the four landraces) resulting from biostimulant applications was 6.64 t ha^{-1} compared to the control. Therefore, the positive effect associated with biostimulant treated plants resulted in an added gross return on San Marzano tomato value of 3320.0 € ha^{-1} (Figure 1). The total added variable cost associated with biostimulant applications was 1956.8 € ha^{-1} and was related particularly to hand harvesting of the added-tomato yield resulting in biostimulant-treated plants. After accounting for added variable costs, the net return of biostimulant-treated compared to untreated control plants was 1363.2 € ha^{-1} (Figure 1). As expected, the shape index was not altered by the growing conditions. Moreover, the PE significantly increased the citric acid in all the landraces but SM2, whose amount linearly correlated with a lower pH value of the fruit juice (Pearson Correlation: -0.492; $p = 0.015$). Acidity is an important factor for tomato flavor, although the limited difference in pH is not expected to considerably influence the suitability for tomato processing [34].

Table 1. Yield, shape index, dry matter percentage, pH and citrate content of the fruits in relation to the landraces (SM1, SM2, SM3, SM4) and the biostimulation application. All data are expressed as mean ± SE, $n = 3$.

Source of Variance	Yield (t ha^{-1})		Shape Index		Dry Matter (%)		pH		Citrate (g kg^{-1} DW)	
Landrace (L)										
SM1	46.68	±2.73 a	1.68	±0.09 b	5.71	±0.09	4.31	±0.02 bc	64.91	±3.30 b
SM2	36.62	±2.10 ab	2.09	±0.07 a	5.66	±0.11	4.35	±0.03 ab	73.57	±2.60 a
SM3	47.62	±4.23 a	2.20	±0.11 a	5.34	±0.22	4.23	±0.04 c	74.11	±3.37 a
SM4	31.10	±3.14 b	1.79	±0.12 b	5.83	±0.10	4.42	±0.03 a	47.24	±1.21 c
Biostimulant (B)										
Control	37.18	±2.30	1.94	±0.23	5.53	±0.12	4.34	±0.03	60.97	±3.60
Plant Extract	43.82	±2.06	1.95	±0.26	5.74	±0.08	4.32	±0.02	68.95	±3.54
Significance										
Landrace (L)	**		***		ns		**		***	
Biostimulant (B)	*		ns		ns		ns		**	
L × B	ns		ns		ns		ns		ns	

ns, *, **, *** Nonsignificant or significant at $p \leq 0.05$, 0.01, and 0.001, respectively. Different letters within each column indicate significant differences according to Duncan's multiple-range test ($p < 0.05$). The factor "Biostimulant" was compared with the Student's t-test.

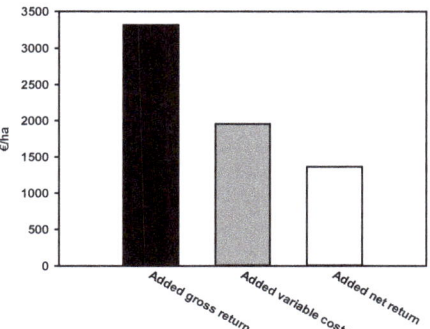

Figure 1. Added returns resulting from biostimulant applications on San Marzano tomato compared to untreated control.

3.2. The Mineral Profile of Fruits Is Mainly Affected by Either the Genotype or Biostimulation

Inorganic ions represent a small fraction of the fruits' dry matter. While the latter was unaffected by our experimental factors, mineral concentration in fruits was extensively and specifically altered by either the landraces or the biostimulation (Table 2). Two mineral elements (P and Ca) were not influenced by the two factors under investigation and their interaction. Nitrate, potassium, and magnesium concentrations varied according to the landrace and biostimulant application (Table 2). The SM3 landraces had a substantially higher content of these elements; for instance, almost double SM1 for nitrate. The effect of biostimulation was mineral-specific because it increased the fruit concentration of the two cations (Mg and K) and decreased that of the NO_3 anion (Table 2). In addition to improving the nutritional value of the fruits (these elements being essential minerals for mammals), their higher accumulation indicates an enhanced mineral utilization efficiency. This is important especially for potassium, because of the possible reduction in the utilization rate of a chemical fertilizer that is required in higher quantity to produce whole peeled tomatoes [35]. While all these alterations were not influenced by the landrace factor, SM4 had a lower sulphate and sodium concentration compared to SM1, SM2 and SM3, irrespective of biostimulant treatment. Finally, mineral composition of the fruits was not shaped by the interaction of the genotype and the biostimulatory treatment. Overall, the data indicated that PE biostimulation can appreciably influence the nutritional value of tomato in a mineral-specific way. It was reported that an algal preparation specifically altered the mineral composition of cherry tomato fruits [36]. Our data indicate that for the different elements, either the landrace or PE biostimulation has a predominant effect. Relative variation in mineral composition was modest, except for nitrate. Anionic or cationic antagonisms were not evident, pointing towards a biostimulatory effect that should be also dependent on mineral transport to the fruits rather than exclusively influencing plant-soil interaction.

3.3. The Sugars and Bioactive Metabolites of the Fruits Are Specifically Changed by Biostimulation

Sugars are the predominant soluble solids of tomatoes and key contributors to their flavor [37]. The starch content in fruits was clearly different among varieties (CV: 23.8%). PE biostimulation had a clear positive effect (+73%) only for one landrace (SM3) while it does not significantly affect the other three (Table 3). Among the main free sugars, sucrose, and fructose, but not glucose, differed among landraces, with the SM3 having the highest content of these saccharides. The PE treatment had a positive effect on fructose and glucose (+13.6% and +26.9%, respectively), similarly to what has been reported in pepper, another Solanaceae [38]. The data indicated that the selected landraces have distinct accumulation patterns of sugars in mature fruits, specifically affected by biostimulation. Anthocyanins, and more generally polyphenols, did not vary according to the experimental factors.

Differences among landraces were present in the lycopene concentration, and this variable was significantly increased (20.8%) by the biostimulation treatment. Even considering the limits of a comparison between different experimental works, this increase is higher than that observed with organic fertilizers [39]. In addition to the known beneficial effects for human health, the improvement of the lycopene content in mature fruits is an important trait for the processing industry, because of the resulting increase of the fruits' red color intensity. Finally, for all these variables but starch, the effect of the biostimulant, when present, was not dependent on the landrace (Table 3).

3.4. Soluble Proteins and Free Amino Acids Profiling Were Principally Affected by the Landrace and Its Interaction with the Biostimulant

Both the variety and its interaction with the biostimulant significantly changed the soluble protein content. Specifically, the SM3 landraces had a significantly higher content of soluble proteins irrespective of the biostimulation. Moreover, in this landrace biostimulation strongly increased the soluble proteins content to a level that was higher than any other experimental conditions. Conversely, the treated SM1, SM2 and SM4 did not show a statistically significant difference from the control plants. Overall, PE biostimulation did not play a significant role in altering the nutritional value of the fruits in terms of total protein amount and free amino acids (AAs), including the essential AAs, except in SM3. This landrace turned out to be the most valuable genetic material along with SM2, which had the highest concentration of total AAs and essential AAs (among which BCAAs), while the lowest values were recorded for SM4. These parameters were not altered by the biostimulation (and factors' interaction). The amount of the AAs was more dependent on the genotype factor (13 AAs significantly affected) than on the biostimulation (six AAs), and factors' interaction (eight AAs). The effect of the PE biostimulation was overall positive, with four (respectively two) AAs present in higher (resp. lower) concentration. Compared to the other landraces, the SM1 had often a reduced quantity of free AAs and it is notable that proline was significantly higher, while other two other stress related AAs, GABA and monoethanolamine (MEA, had quantities similar to those of the other landraces. In the fruits of biostimulated SM1, total AAs content increased by about 50% compared to the respective control. Interestingly, biostimulation halved the amount of proline, while not affecting GABA.

Irrespective of the treatment, glutamate, glutamine, GABA, asparagine, and aspartate were the most abundant AAs, representing about 26.7%, 23.9%, 16.0%, 12.8% and 5.3% of total free AAs in all samples (Table 4). This profile is in agreement with that of the San Marzano varieties [40]. Therefore, it may be not casual that significant variations between the different SM landraces were observed for other AAs, such as arginine (SM2, SM3), MEA (SM1, SM2), lysine (SM2) and ornithine (SM2, SM4), which, along with GABA (SM1, SM2, SM3), were significantly higher in the indicated SM landraces compared to the other ones. Averaged over the different landraces, biostimulation significantly increased alanine and glycine concentration by 44.6% and 35.7%, respectively, compared to untreated plants. The highest asparagine content was present in SM3 that was equal to 8.34 ± 1.06 µmol g^{-1} FW. Biostimulant application increased the content of this AA only in SM1 (+2.7-fold compared to the untreated control). SM2 and SM3 had a glutamine content (14.7 µmol g^{-1} FW on average) higher than other two landraces. SM1 under control conditions had the highest proline of the four landraces, that was equal to 0.60 ± 0.08 µmol g^{-1} FW but decreased following biostimulation. SM2 had the highest content of essential AAs, which was on average 4.13 µmol g^{-1} FW. Only in the treated-SM1 fruits, essential AAs increased compared to the corresponding control, while no differences were observed in the other SME. The same trend was observed for the BCAAs content (Table 4).

3.5. Princpial Component Analysis Indicated the Complex Relationships of the Biostimulatory Response

To highlight a possible underlying structure of the dataset, we summarized and visualized the various characteristics of the samples through multivariate analyses. Specifically, we used a Principal Component Analysis (PCA) to highlight patterns of variation from our range of different categories of measurements (Figure 2).

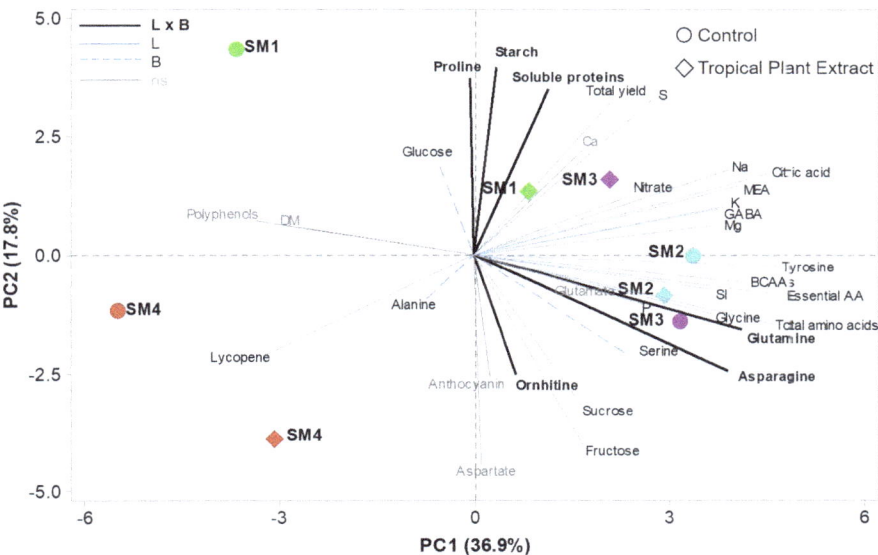

Figure 2. Principal Component Analysis bi plot. The first two PCs explained 54.7% of the total variance. Tropical plant extract-biostimulated landraces are represented by squares, control landraces by circles. Scaled loading vectors are drawn as black bold continuous line for variables that were signicantly affected by both the landrace and the biostimulant factors, blue continuous line for those significantly affected by the landraces, dashed blue line for those signifcanty affected by the biostimulant (only one occurrence) and as a thin gray line for variables that were not significantly influenced.

The PCA analysis indicated that the first four PCs had eigen values higher than one and explained 81.3% of the total variance, with PC1, PC2, PC3 and PC4 accounting for 36.9%, 17.8%, 16.0% and 10.5%, respectively. The first principal component had large positive associations with many AAs (e.g., tyrosine, essential AAs. BCAAs, amides and MEA), in addition to citric acid, and it was negatively correlated with lycopene (Figure 2). PC2 had a strong positive association with proline, starch, and soluble proteins content, while it was negatively correlated to fructose and sucrose content. Moreover, the loading plot indicated that proline content poorly correlated with the other AAs. The distribution of landraces based on the first two PCs indicated that trait variation was ample and able to disperse the samples along the two axes. Specifically, the SM1 and SM4 controls were present in distinct positions (i.e., SM1 in the upper and SM4 close to the negative side of PC1 in the lower left quadrant). The application of biostimulant strongly changed the distribution of these two landraces mainly along PC1 for SM1, and PC2 for SM4. Conversely, SM2 and SM3, present in the lower right quadrant close to x-axis, were primarily separated along the second axis, while the treated SM2 remained closer to its control and the treated SM3 moved in the upper right quadrant. Factor loadings and biplot analysis indicated that several, mostly uncorrelated, traits were the major determinants of observed diversity. The landrace scatter plot was not able to depict a clear pattern of grouping according to the genotype or the treatment. Overall, the multivariate analyses highlighted the biodiversity of tomato landraces (with similar geographical origin and destination of use) with respect to physicochemical, nutritional, and functional traits, which is evident also considering their different response to the PE biostimulant in non-stress conditions.

Table 2. Mineral composition of the tomato fruits in relation to the landraces (SM1, SM2, SM3, SM4) and the biostimulation application. All data are expressed as mean ± SE, $n = 3$.

Source of Variance	NO$_3$-N (mg kg^{-1} FW)	P (g kg^{-1} DW)	K (g kg^{-1} DW)	Ca (g kg^{-1} DW)	Mg (g kg^{-1} DW)	S (g kg^{-1} DW)	Na (g kg^{-1} DW)
Landrace (L)							
SM1	77.71 ±5.78 ab	3.20 ±0.18	49.21 ±2.10 ab	0.96 ±0.05	1.97 ±0.05 b	0.84 ±0.04 a	0.26 ±0.02 a
SM2	67.75 ±3.85 bc	3.26 ±0.20	53.55 ±1.32 ab	0.96 ±0.05	2.30 ±0.10 a	0.83 ±0.03 a	0.28 ±0.02 a
SM3	91.05 ±9.92 a	3.18 ±0.23	57.93 ±5.38 a	0.85 ±0.05	2.34 ±0.15 a	0.77 ±0.06 a	0.24 ±0.02 a
SM4	51.74 ±6.03 c	3.03 ±0.21	44.47 ±0.69 b	0.81 ±0.06	1.89 ±0.10 b	0.64 ±0.04 b	0.17 ±0.01 b
Biostimulant (B)							
Control	80.04 ±6.92	3.21 ±0.16	48.10 ±1.31	0.93 ±0.04	2.02 ±0.08	0.73 ±0.04	0.24 ±0.02
Tropical plant extract	64.08 ±4.42	3.13 ±0.12	54.48 ±3.01	0.86 ±0.04	2.22 ±0.09	0.80 ±0.03	0.23 ±0.01
Significance							
Landrace (L)	**	ns	*	ns	**	**	***
Biostimulant (B)	*	ns	*	ns	*	ns	ns
L × B	ns	ns	ns	ns	ns	ns	ns

ns, *, **, *** Nonsignificant or significant at $p \leq 0.05$, 0.01, and 0.001, respectively. Different letters within each column indicate significant differences according to Duncan's multiple-range test ($p < 0.05$). The factor "Biostimulant" was compared with the Student's t-test.

Table 3. Starch, free sugars, anthocyanins, lycopene, and polyphenols in relation to the landraces (SM1, SM2, SM3, SM4) and the biostimulation application. All data are expressed as mean ± SE, $n = 3$.

Source of Variance	Starch (μmol g^{-1} FW)	Glucose (μmol g^{-1} FW)	Fructose (μmol g^{-1} FW)	Sucrose (μmol g^{-1} FW)	Anthocyanins (mg C3G 100 g^{-1} FW)	Lycopene (mg 100 g^{-1} FW)	Polyphenols (mg GAE 100 g^{-1} FW)
Landrace (L)							
SM1	31.99 ±2.58 a	64.49 ±1.56	14.18 ±2.75 b	21.73 ±2.49 b	44.60 ±2.12	15.73 ±2.15 ab	37.03 ±1.98
SM2	20.48 ±1.04 c	58.27 ±3.54	28.42 ±2.71 a	23.90 ±2.04 b	50.44 ±2.87	13.47 ±1.22 b	34.97 ±2.35
SM3	26.25 ±3.84 b	63.10 ±5.74	26.61 ±2.22 a	29.87 ±2.31 a	47.50 ±2.06	11.78 ±1.18 b	36.10 ±3.47
SM4	19.28 ±0.25 c	60.88 ±2.45	27.76 ±1.23 a	27.12 ±1.21 ab	49.93 ±3.90	19.53 ±1.83 a	38.70 ±2.15
Biostimulant (B)							
Control	22.68 ±1.34	57.76 ±2.60 b	21.37 ±2.60	23.92 ±2.03	47.56 ±1.44	13.70 ±1.27	36.20 ±1.76
Tropical plant extract	26.32 ±2.72	65.61 ±1.92 a	27.12 ±1.68	27.39 ±1.00	48.68 ±2.48	16.55 ±1.42	37.21 ±1.78
Significance							
Landrace (L)	***	ns	***	*	ns	**	ns
Biostimulant (B)	ns	*	*	ns	ns	*	ns
L × B	*	ns	ns	ns	ns	ns	ns

ns, *, **, *** Nonsignificant or significant at $p \leq 0.05$, 0.01, and 0.001, respectively. Different letters within each column indicate significant differences according to Duncan's multiple-range test ($p < 0.05$). The factor "Biostimulant" was compared with the Student's t-test.

Table 4. Amino acidic profile of the fruits in relation to the landraces (SM1, SM2, SM3, SM4) and the biostimulation applications. All data are expressed as mean ± standard error, $n = 3$.

Chemical Compound	Landrace				sig	Biostimulant		sig	Landrace × Biostimulant								sig
	SM1	SM2	SM3	SM4		Control (C)	Biostimulation (PE)		SM1 C	SM2 C	SM3 C	SM4 C	SM1 PE	SM2 PE	SM3 PE	SM4 PE	
Soluble proteins [a]	26.15 ± 2.29 ab	22.75 ± 1.49 bc	30.05 ± 3.48 a	20.05 ± 1.82 c	**	24.49 ± 1.32	25.01 ± 2.45	ns	27.94 ± 3.63 b	25.66 ± 1.36 bc	22.90 ± 0.38 bc	21.45 ± 3.26 bc	24.36 ± 3.15 bc	19.83 ± 0.87 bc	37.20 ± 3.04 a	18.65 ± 1.98 c	**
Alanine [b]	1.22 ± 0.13	1.07 ± 0.15	0.68 ± 0.07	1.11 ± 0.26	ns	0.83 ± 0.11	1.20 ± 0.12	*	1.14 ± 0.26	0.95 ± 0.26	0.65 ± 0.10	0.59 ± 0.15	1.29 ± 0.11	1.18 ± 0.16	0.71 ± 0.13	1.63 ± 0.20	ns
Arginine [b]	0.85 ± 0.05 b	1.55 ± 0.15 a	1.35 ± 0.20 a	0.55 ± 0.06 b	***	1.06 ± 0.16	1.08 ± 0.13	ns	0.74 ± 0.03	1.56 ± 0.08	1.49 ± 0.32	0.46 ± 0.11	0.95 ± 0.04	1.54 ± 0.31	1.21 ± 0.27	0.63 ± 0.02	ns
Asparagine [b]	4.41 ± 0.92 c	6.76 ± 0.50 ab	8.34 ± 1.06 a	5.18 ± 0.72 bc	*	5.83 ± 0.95	6.52 ± 0.35	ns	2.43 ± 0.41 c	7.02 ± 0.93 ab	9.52 ± 1.84 a	4.35 ± 0.92 bc	6.39 ± 0.34 ab	6.49 ± 0.55 ab	7.17 ± 0.91 ab	6.02 ± 1.04 b	*
Aspartate [b]	2.32 ± 0.27	2.44 ± 0.20	2.59 ± 0.33	2.87 ± 0.38	ns	2.37 ± 0.24	2.74 ± 0.17	ns	1.90 ± 0.13	2.17 ± 0.25	2.85 ± 0.54	2.56 ± 0.76	2.75 ± 0.42	2.70 ± 0.26	2.33 ± 0.44	3.17 ± 0.22	ns
MEA [b]	0.10 ± 0.01 a	0.12 ± 0.01 a	0.09 ± 0.01 ab	0.07 ± 0.01 b	*	0.09 ± 0.01	0.09 ± 0.01	ns	0.09 ± 0.01	0.11 ± 0.01	0.10 ± 0.01	0.07 ± 0.01	0.11 ± 0.02	0.12 ± 0.03	0.09 ± 0.01	0.06 ± 0.01	ns
GABA [b]	7.89 ± 0.92 a	10.33 ± 1.43 a	7.95 ± 0.91 a	4.84 ± 0.50 b	**	7.21 ± 0.67	8.29 ± 1.04	ns	7.05 ± 1.53	8.33 ± 0.62	8.55 ± 1.51	4.92 ± 0.90	8.73 ± 1.08	12.33 ± 2.42	7.34 ± 1.22	4.75 ± 0.67	ns
Glycine [b]	0.12 ± 0.01 b	0.20 ± 0.02 a	0.20 ± 0.03 a	0.14 ± 0.02 b	**	0.14 ± 0.02	0.19 ± 0.02	*	0.10 ± 0.01	0.21 ± 0.02	0.16 ± 0.02	0.10 ± 0.02	0.14 ± 0.01	0.20 ± 0.03	0.24 ± 0.05	0.18 ± 0.03	ns
Glutamate [b]	13.75 ± 1.79	13.66 ± 1.87	12.45 ± 1.89	11.81 ± 1.40	ns	12.01 ± 1.21	13.82 ± 1.29	ns	11.42 ± 1.21	10.57 ± 0.45	14.24 ± 3.19	11.81 ± 2.83	16.09 ± 3.01	16.74 ± 2.78	10.66 ± 2.11	11.81 ± 1.33	ns
Glutamine [b]	7.94 ± 1.21 b	13.76 ± 1.79 a	15.64 ± 1.18 a	8.88 ± 1.33 b	***	11.51 ± 1.71	11.60 ± 0.87	ns	5.77 ± 0.50 d	16.72 ± 1.84 a	16.83 ± 1.53 a	6.74 ± 1.64 cd	10.12 ± 1.55 bcd	10.80 ± 1.96 bcd	14.45 ± 1.80 ab	11.02 ± 1.25 bc	***
Isoleucine [b]	0.43 ± 0.05	0.57 ± 0.06	0.42 ± 0.04	0.36 ± 0.04	ns	0.41 ± 0.04	0.49 ± 0.04	ns	0.34 ± 0.04	0.53 ± 0.10	0.42 ± 0.05	0.33 ± 0.08	0.53 ± 0.06	0.60 ± 0.09	0.43 ± 0.07	0.39 ± 0.00	ns
Histidine [b]	0.43 ± 0.06 bc	0.61 ± 0.04 a	0.53 ± 0.07 ab	0.35 ± 0.04 c	**	0.47 ± 0.05	0.49 ± 0.04	ns	0.35 ± 0.06	0.61 ± 0.03	0.62 ± 0.11	0.31 ± 0.07	0.51 ± 0.08	0.61 ± 0.09	0.44 ± 0.05	0.38 ± 0.04	ns
Leucine [b]	0.37 ± 0.04 b	0.48 ± 0.02 a	0.33 ± 0.02 b	0.32 ± 0.03 c	**	0.35 ± 0.03	0.40 ± 0.03	ns	0.30 ± 0.03	0.47 ± 0.02	0.36 ± 0.02	0.27 ± 0.04	0.43 ± 0.07	0.49 ± 0.03	0.31 ± 0.04	0.36 ± 0.04	ns
Lysine [b]	0.40 ± 0.06 b	0.60 ± 0.05 a	0.43 ± 0.04 b	0.35 ± 0.05 b	**	0.44 ± 0.03	0.46 ± 0.03	ns	0.29 ± 0.04 d	0.67 ± 0.06 a	0.49 ± 0.08 abc	0.31 ± 0.06 cd	0.52 ± 0.04 ab	0.54 ± 0.05 ab	0.38 ± 0.02 bcd	0.39 ± 0.09 bcd	*
Methionine [b]	0.07 ± 0.01 bc	0.11 ± 0.00 a	0.08 ± 0.01 b	0.05 ± 0.01 c	***	0.07 ± 0.01	0.09 ± 0.01	*	0.05 ± 0.00 de	0.12 ± 0.01 a	0.08 ± 0.01 cd	0.04 ± 0.01 e	0.09 ± 0.01 bc	0.11 ± 0.00 ab	0.08 ± 0.01 cd	0.07 ± 0.01 cd	*
Ornithine [b]	0.20 ± 0.03 b	0.32 ± 0.06 a	0.20 ± 0.02 b	0.30 ± 0.03 a	**	0.27 ± 0.03	0.25 ± 0.02	ns	0.16 ± 0.00	0.44 ± 0.06 a	0.21 ± 0.02 c	0.25 ± 0.01 bc	0.24 ± 0.04 c	0.20 ± 0.03 c	0.20 ± 0.02 c	0.35 ± 0.04 ab	***
Phenylalanine [b]	0.68 ± 0.09	0.85 ± 0.09	0.66 ± 0.09	0.68 ± 0.11	ns	0.63 ± 0.05	0.80 ± 0.07	*	0.54 ± 0.08	0.73 ± 0.04	0.69 ± 0.15	0.55 ± 0.13	0.81 ± 0.11	0.97 ± 0.15	0.62 ± 0.13	0.81 ± 0.17	ns
Proline [b]	0.40 ± 0.10 a	0.30 ± 0.03 a	0.20 ± 0.03 b	0.11 ± 0.00 b	***	0.31 ± 0.06	0.19 ± 0.03	**	0.60 ± 0.08 a	0.28 ± 0.05 bc	0.25 ± 0.02 bcd	0.12 ± 0.00 de	0.20 ± 0.02 bcde	0.32 ± 0.05 b	0.15 ± 0.03 cde	0.10 ± 0.01 e	***
Serine [b]	0.90 ± 0.10	1.18 ± 0.04	1.06 ± 0.16	0.99 ± 0.27	ns	0.82 ± 0.09	1.25 ± 0.10	**	0.72 ± 0.11	1.16 ± 0.01	0.91 ± 0.16	0.49 ± 0.12	1.08 ± 0.07	1.20 ± 0.10	1.21 ± 0.28	1.49 ± 0.29	ns
Tyrosine [b]	0.22 ± 0.02 b	0.38 ± 0.03 a	0.34 ± 0.02 a	0.18 ± 0.02 b	***	0.26 ± 0.03	0.30 ± 0.03	ns	0.19 ± 0.03	0.36 ± 0.06	0.34 ± 0.03	0.16 ± 0.03	0.25 ± 0.02	0.40 ± 0.04	0.34 ± 0.05	0.20 ± 0.04	ns
Threonine [b]	0.28 ± 0.04	0.34 ± 0.08	0.38 ± 0.08	0.23 ± 0.02	ns	0.29 ± 0.05	0.33 ± 0.04	ns	0.21 ± 0.03	0.28 ± 0.10	0.43 ± 0.14	0.22 ± 0.04	0.35 ± 0.06	0.40 ± 0.14	0.32 ± 0.11	0.24 ± 0.03	ns
Tryptophan [b]	0.09 ± 0.01 c	0.16 ± 0.01 ab	0.17 ± 0.03 a	0.13 ± 0.01 b	***	0.15 ± 0.02	0.13 ± 0.01	ns	0.08 ± 0.00 d	0.17 ± 0.02 b	0.23 ± 0.01 a	0.14 ± 0.02 bc	0.11 ± 0.01 cd	0.15 ± 0.01 bc	0.12 ± 0.01 c	0.13 ± 0.01 bc	***
Valine [b]	0.23 ± 0.04 c	0.41 ± 0.08 ab	0.31 ± 0.02 b	0.18 ± 0.02 c	***	0.27 ± 0.04	0.30 ± 0.02	ns	0.17 ± 0.02 d	0.45 ± 0.03 a	0.30 ± 0.02 bc	0.15 ± 0.04 d	0.30 ± 0.04 bc	0.37 ± 0.03 ab	0.32 ± 0.05 b	0.22 ± 0.00 cd	ns
Essential AA [b]	3.84 ± 0.41 bc	5.68 ± 0.36 a	4.67 ± 0.49 ab	3.21 ± 0.30 c	**	4.14 ± 0.42	4.56 ± 0.34	ns	3.07 ± 0.25	5.58 ± 0.23	5.11 ± 0.73	2.80 ± 0.43	4.61 ± 0.44	5.78 ± 0.77	4.23 ± 0.70	3.62 ± 0.30	ns
BCAAs [b]	1.03 ± 0.13 b	1.45 ± 0.07 a	1.07 ± 0.08 b	0.86 ± 0.08 c	***	1.02 ± 0.09	1.19 ± 0.08	ns	0.81 ± 0.08	1.44 ± 0.11	1.08 ± 0.09	0.76 ± 0.13	1.26 ± 0.17	1.46 ± 0.12	1.06 ± 0.15	0.97 ± 0.03	ns
Total amino acids	43.31 ± 4.94 ab	56.20 ± 4.10 a	54.42 ± 5.36 a	39.67 ± 4.42 b	*	45.80 ± 4.31	51.00 ± 3.10	ns	34.63 ± 4.05	53.92 ± 2.64	59.71 ± 8.90	34.94 ± 7.43	51.99 ± 5.51	58.48 ± 8.47	49.13 ± 6.05	44.40 ± 4.49	ns

[a]: mg g^{-1} FW; [b]: μmol g^{-1} FW; ns, *, **, *** Nonsignificant or significant at $p \leq 0.05$, 0.01, and 0.001, respectively. Different letters within each row indicate significant differences according to Duncan's multiple-range test ($p < 0.05$). The factor "Biostimulant" was compared with the Student's t-test.

4. Conclusions

Our work provided evidence on the effect of a plant-based biostimulant on indeterminate, open-pollinated tomatoes that have been largely developed through adaptation rather than formal breeding. The targeted phytochemical profile indicated the presence of substantial variation among the landraces and their response to biostimulation, which proved to be capable of inducing landrace-specific beneficial features to yield and fruit quality. This observation would be consistent with a biostimulatory activity that is not acting towards one or few specific plant functions [41]. Under this perspective, our work showed that in tomato landraces, the influence of the tropical PE on the chemical and biochemical fruit composition is characterized by a substantial flexibility, in terms of both magnitude and type of altered parameters. Moreover, this response has a variable degree of correlation with the plant genotype, which may account for the sometimes-contrasting reports on biostimulation in tomato, and other species, under non-stress conditions [17,42]. The data also underlined our partial capacity to model fruit quality in response to biostimulation, an area that deserves further integrative investigations. Considering each single landrace, the scale of the modification (e.g., on dry weight, acidity, starch, free amino acids) implies that the tropical plant extract is not expected to affect largely the attributes of the berries that are important for their traditional destination of use (i.e., whole-peeled tomatoes). Interestingly, a positive impact of the biostimulation was present for some taste-related, nutritional, and functional quality traits of the fruits (e.g., simple sugars, lycopene, some organic acids and macroelements).

Supplementary Materials: The following are available online at https://www.mdpi.com/article/10.3390/foods10050926/s1, Supplementary Table S1. Morphological characteristics of the four SM landraces. Quantitative traits are reported as mean ± standard error, n = 3.

Author Contributions: Conceptualization Y.R., G.C. (Giuseppe Colla); Formal analysis G.C. (Giandomenico Corrado) and P.C.; Funding acquisition: P.C.; Investigation; E.D., L.I.D., G.M.F.; Methodology: E.D., L.I.D.; Supervision Y.R. and G.C. (Giandomenico Corrado); Roles/Writing—original draft: Y.R., G.C. (Giandomenico Corrado), and P.C.; Writing—review & editing: Y.R., G.C. (Giandomenico Corrado), S.D.P., G.C. (Giuseppe Colla), and P.C. All authors have read and agreed to the published version of the manuscript.

Funding: This research received no external funding.

Data Availability Statement: The datasets generated for this study are available on request to the corresponding author.

Acknowledgments: This work was funded by Regione Campania Lotta alle Patologie Oncologiche progetto iCURE (CUP B21C17000030007—SURF 17061BP000000008). The authors are grateful to Christophe El-Nakhel, Beniamino Gentile, and Giampaolo Raimondi for their assistance in the greenhouse experiment, and to Antonio Pannico for his assistance in the laboratory analyses.

Conflicts of Interest: The authors declare no conflict of interest.

References

1. Martínez-Carrasco, L.; Brugarolas, M.; Martínez-Poveda, A.; Ruiz, J.J.; García-Martínez, S. Modelling perceived quality of tomato by structural equation analysis. *Br. Food J.* **2012**, *114*, 1414–1431. [CrossRef]
2. Marsden, T.; Morley, A. *Sustainable Food Systems: Building a New Paradigm*; Marsden, T., Morley, A., Eds.; Routledge: London, UK, 2014; ISBN 9780203083499.
3. Allen, T.; Prosperi, P.; Cogill, B.; Flichman, G. Agricultural biodiversity, social-ecological systems and sustainable diets. *Proc. Nutr. Soc.* **2014**, *73*, 498–508. [CrossRef]
4. Bell, J.; Paula, L.; Dodd, T.; Németh, S.; Nanou, C.; Mega, V.; Campos, P. EU ambition to build the world's leading bioeconomy—Uncertain times demand innovative and sustainable solutions. *New Biotechnol.* **2018**, *40*, 25–30. [CrossRef]
5. Dwivedi, S.; Goldman, I.; Ortiz, R. Pursuing the potential of heirloom cultivars to improve adaptation, nutritional, and culinary features of food crops. *Agronomy* **2019**, *9*, 441. [CrossRef]
6. Bauchet, G.; Causse, M. Genetic Diversity in Tomato (*Solanum lycopersicum*) and Its Wild Relatives. *Genet. Divers. Plants* **2012**, *8*, 134–162. [CrossRef]

7. Digilio, M.C.; Corrado, G.; Sasso, R.; Coppola, V.; Iodice, L.; Pasquariello, M.; Bossi, S.; Maffei, M.E.; Coppola, M.; Pennacchio, F.; et al. Molecular and chemical mechanisms involved in aphid resistance in cultivated tomato. *New Phytol.* **2010**, *187*, 1089–1101. [CrossRef]
8. Figàs, M.R.; Prohens, J.; Raigón, M.D.; Fita, A.; García-Martínez, M.D.; Casanova, C.; Borràs, D.; Plazas, M.; Andújar, I.; Soler, S. Characterization of composition traits related to organoleptic and functional quality for the differentiation, selection and enhancement of local varieties of tomato from different cultivar groups. *Food Chem.* **2015**, *187*, 517–524. [CrossRef]
9. Massaretto, I.L.; Albaladejo, I.; Purgatto, E.; Flores, F.B.; Plasencia, F.; Egea-Fernández, J.M.; Bolarin, M.C.; Egea, I. Recovering tomato landraces to simultaneously improve fruit yield and nutritional quality against salt stress. *Front. Plant Sci.* **2018**, *871*, 1778. [CrossRef] [PubMed]
10. Corrado, G. Advances in DNA typing in the agro-food supply chain. *Trends Food Sci. Technol.* **2016**, *52*, 80–89. [CrossRef]
11. Corrado, G.; Imperato, A.; la Mura, M.; Perri, E.; Rao, R. Genetic diversity among olive varieties of southern italy and the traceability of olive oil using SSR markers. *J. Hortic. Sci. Biotechnol.* **2011**, *86*, 461–466. [CrossRef]
12. Scarano, D.; Rao, R.; Masi, P.; Corrado, G. SSR fingerprint reveals mislabeling in commercial processed tomato products. *Food Control* **2015**, *51*, 397–401. [CrossRef]
13. Du Jardin, P. Plant biostimulants: Definition, concept, main categories and regulation. *Sci. Hortic.* **2015**, *196*, 3–14. [CrossRef]
14. Rouphael, Y.; Colla, G. Biostimulants in Agriculture. *Front. Plant Sci.* **2020**, *11*. [CrossRef] [PubMed]
15. Colla, G.; Hoagland, L.; Ruzzi, M.; Cardarelli, M.; Bonini, P.; Canaguier, R.; Rouphael, Y. Biostimulant action of protein hydrolysates: Unraveling their effects on plant physiology and microbiome. *Front. Plant Sci.* **2017**, *8*, 2202. [CrossRef] [PubMed]
16. Ertani, A.; Pizzeghello, D.; Francioso, O.; Sambo, P.; Sanchez-Cortes, S.; Nardi, S. Capsicum chinensis L. growth and nutraceutical properties are enhanced by biostimulants in a long-term period: Chemical and metabolomic approaches. *Front. Plant Sci.* **2014**, *5*, 375. [CrossRef]
17. Rodrigues, M.; Baptistella, J.L.C.; Horz, D.C.; Bortolato, L.M.; Mazzafera, P. Organic plant biostimulants and fruit quality-a review. *Agronomy* **2020**, *10*, 988. [CrossRef]
18. Colla, G.; Cardarelli, M.; Bonini, P.; Rouphael, Y. Foliar applications of protein hydrolysate, plant and seaweed extracts increase yield but differentially modulate fruit quality of greenhouse tomato. *HortScience* **2017**, *52*, 1214–1220. [CrossRef]
19. Colla, G.; Rouphael, Y. Biostimulants in horticulture. *Sci. Hortic.* **2015**, *196*, 1–2. [CrossRef]
20. Rouphael, Y.; Giordano, M.; Cardarelli, M.; Cozzolino, E.; Mori, M.; Kyriacou, M.C.; Bonini, P.; Colla, G. Plant-and seaweed-based extracts increase yield but differentially modulate nutritional quality of greenhouse spinach through biostimulant action. *Agronomy* **2018**, *8*, 126. [CrossRef]
21. Xu, L.; Geelen, D. Developing biostimulants from agro-food and industrial by-products. *Front. Plant Sci.* **2018**, *871*, 1567. [CrossRef]
22. Caramante, M.; Rao, R.; Monti, L.M.; Corrado, G. Discrimination of "San Marzano" accessions: A comparison of minisatellite, CAPS and SSR markers in relation to morphological traits. *Sci. Hortic.* **2009**, *120*, 560–564. [CrossRef]
23. Monti, L.; Santangelo, E.; Corrado, G.; Rao, R.; Soressi, G.P.; Scarascia Mugnozza, G.T. Il "SanMarzano": Problematiche e prospettive in relazione alla sua salvaguardia e alla necessità di interventi genetici. *Agroindustria* **2004**, *3*, 161–169.
24. Rouphael, Y.; Colla, G.; Giordano, M.; El-Nakhel, C.; Kyriacou, M.C.; De Pascale, S. Foliar applications of a legume-derived protein hydrolysate elicit dose-dependent increases of growth, leaf mineral composition, yield and fruit quality in two greenhouse tomato cultivars. *Sci. Hortic.* **2017**, *226*, 353–360. [CrossRef]
25. Carillo, P.; Colla, G.; El-Nakhel, C.; Bonini, P.; D'Amelia, L.; Dell'Aversana, E.; Pannico, A.; Giordano, M.; Sifola, M.I.; Kyriacou, M.C.; et al. Biostimulant application with a tropical plant extract enhances corchorus olitorius adaptation to sub-optimal nutrient regimens by improving physiological parameters. *Agronomy* **2019**, *9*, 249. [CrossRef]
26. Carillo, P.; Kyriacou, M.C.; El-Nakhel, C.; Pannico, A.; dell'Aversana, E.; D'Amelia, L.; Colla, G.; Caruso, G.; De Pascale, S.; Rouphael, Y. Sensory and functional quality characterization of protected designation of origin 'Piennolo del Vesuvio' cherry tomato landraces from Campania-Italy. *Food Chem.* **2019**, *292*, 166–175. [CrossRef]
27. Mónica Giusti, M.; Wrolstad, R.E. Characterization and Measurement of Anthocyanins by UV-visible Spectroscopy. *Handb. Food Anal. Chem.* **2005**, *2*, 19–31. [CrossRef]
28. Sadler, G.; Davis, J.; Dezman, D. Rapid Extraction of Lycopene and β-Carotene from Reconstituted Tomato Paste and Pink Grapefruit Homogenates. *J. Food Sci.* **1990**, *55*, 1460–1461. [CrossRef]
29. Singleton, V.L.; Orthofer, R.; Lamuela-Raventós, R.M. Analysis of total phenols and other oxidation substrates and antioxidants by means of folin-ciocalteu reagent. *Methods Enzymol.* **1999**, *299*, 152–178. [CrossRef]
30. Carillo, P.; Cacace, D.; De Rosa, M.; De Martino, E.; Cozzolino, C.; Nacca, F.; D'Antonio, R.; Fuggi, A. Process optimisation and physicochemical characterisation of potato powder. *Int. J. Food Sci. Technol.* **2009**, *44*, 145–151. [CrossRef]
31. Woodrow, P.; Ciarmiello, L.F.; Annunziata, M.G.; Pacifico, S.; Iannuzzi, F.; Mirto, A.; D'Amelia, L.; Dell'Aversana, E.; Piccolella, S.; Fuggi, A.; et al. Durum wheat seedling responses to simultaneous high light and salinity involve a fine reconfiguration of amino acids and carbohydrate metabolism. *Physiol. Plant.* **2017**, *159*, 290–312. [CrossRef]
32. Djidonou, D.; Gao, Z.; Zhao, X. Economic Analysis of Grafted Tomato Production in Sandy Soils in Northern Florida. *HortTechnology* **2013**, *23*, 613. [CrossRef]
33. Metsalu, T.; Vilo, J. ClustVis: A web tool for visualizing clustering of multivariate data using Principal Component Analysis and heatmap. *Nucleic Acids Res.* **2015**, *43*, W566–W570. [CrossRef] [PubMed]

34. Young, T.E.; Juvik, J.A.; Sullivan, J.G. Accumulation of the Components of Total Solids in Ripening Fruits of Tomato. *J. Am. Soc. Hortic. Sci.* **1993**, *118*, 286–292. [CrossRef]
35. Hartz, T.K.; Miyao, G.; Mullen, R.J.; Cahn, M.D.; Valencia, J.; Brittan, K.L. Potassium requirements for maximum yield and fruit quality of processing tomato. *J. Am. Soc. Hortic. Sci.* **1999**, *124*, 199–204. [CrossRef]
36. Dobromilska, R.; Mikiciuk, M.; Gubarewicz, K. Evaluation of cherry tomato yielding and fruit mineral composition after using of Bio-algeen S-90 preparation. *J. Elem.* **2008**, *13*, 491–499.
37. Kader, A.A. Flavor quality of fruits and vegetables. *J. Sci. Food Agric.* **2008**, *88*, 1863–1868. [CrossRef]
38. Barrajón-Catalán, E.; Álvarez-Martínez, F.J.; Borrás, F.; Pérez, D.; Herrero, N.; Ruiz, J.J.; Micol, V. Metabolomic analysis of the effects of a commercial complex biostimulant on pepper crops. *Food Chem.* **2020**, *310*, 125818. [CrossRef]
39. Bilalis, D.; Krokida, M.; Roussis, I.; Papastylianou, P.; Travlos, I.; Cheimona, N.; Dede, A. Effects of organic and inorganic fertilization on yield and quality of processing tomato (Lycopersicon esculentum Mill.). *Folia Hortic.* **2018**, *30*, 321–332. [CrossRef]
40. Loiudice, R.; Impembo, M.; Laratta, B.; Villari, G.; Lo Voi, A.; Siviero, P.; Castaldo, D. Composition of San Marzano tomato varieties. *Food Chem.* **1995**, *53*, 81–89. [CrossRef]
41. Drobek, M.; Frąc, M.; Cybulska, J. Plant biostimulants: Importance of the quality and yield of horticultural crops and the improvement of plant tolerance to abiotic stress-a review. *Agronomy* **2019**, *9*, 335. [CrossRef]
42. Bulgari, R.; Cocetta, G.; Trivellini, A.; Vernieri, P.; Ferrante, A. Biostimulants and crop responses: A review. *Biol. Agric. Hortic.* **2015**, *31*, 1–17. [CrossRef]

Article

Bacterial Fertilizers Based on *Rhizobium laguerreae* and *Bacillus halotolerans* Enhance *Cichorium endivia* L. Phenolic Compound and Mineral Contents and Plant Development

Alejandro Jiménez-Gómez [1,2,*,†], Ignacio García-Estévez [3], M. Teresa Escribano-Bailón [3], Paula García-Fraile [1,2] and Raúl Rivas [1,2,4]

[1] Departamento de Microbiología y Genética, Universidad de Salamanca, Edificio Departamental de Biología, 37007 Salamanca, Spain; paulagf81@usal.es (P.G.-F.); raulrg@usal.es (R.R.)
[2] Spanish-Portuguese Institute for Agricultural Research (CIALE), 37185 Salamanca, Spain
[3] Grupo de Investigación en Polifenoles (GIP), Departamento de Química Analítica, Nutrición y Bromatología, Faculty of Pharmacy, Universidad de Salamanca, 37007 Salamanca, Spain; igarest@usal.es (I.G.-E.); escriban@usal.es (M.T.E.-B.)
[4] Associated Unit USAL-CSIC (IRNASA), 37008 Salamanca, Spain
* Correspondence: alexjg@usal.es
† Present address: School of Humanities and Social Sciences, University Isabel I, 09003 Burgos, Spain.

Abstract: Today there is an urgent need to find new ways to satisfy the current and growing food demand and to maintain crop protection and food safety. One of the most promising changes is the replacement of chemical fertilizers with biofertilizers, which include plant root-associated beneficial bacteria. This work describes and shows the use of B. halotolerans SCCPVE07 and R. laguerreae PEPV40 strains as efficient biofertilizers for escarole crops, horticultural species that are widely cultivated. An in silico genome study was performed where coding genes related to plant growth promoting (PGP) mechanisms or different enzymes implicated in the metabolism of phenolic compounds were identified. An efficient bacterial root colonization process was also analyzed through fluorescence microscopy. SCCPVE07 and PEPV40 promote plant development under normal conditions and saline stress. Moreover, inoculated escarole plants showed not only an increase in potassium, iron and magnesium content but also a significant improvement in protocatechuic acid, caffeic acid or kaempferol 3-O-glucuronide plant content. Our results show for the first time the beneficial effects in plant development and the food quality of escarole crops and highlight a potential and hopeful change in the current agricultural system even under saline stress, one of the major non-biological stresses.

Keywords: phenolic acids; escarole; bioactive compounds; biofertilizer; flavonols

Citation: Jiménez-Gómez, A.; García-Estévez, I.; Escribano-Bailón, M.T.; García-Fraile, P.; Rivas, R. Bacterial Fertilizers Based on *Rhizobium laguerreae* and *Bacillus halotolerans* Enhance *Cichorium endivia* L. Phenolic Compound and Mineral Contents and Plant Development. *Foods* **2021**, *10*, 424. https://doi.org/10.3390/foods10020424

Academic Editor: Lisa Pilkington

Received: 26 January 2021
Accepted: 12 February 2021
Published: 15 February 2021

Publisher's Note: MDPI stays neutral with regard to jurisdictional claims in published maps and institutional affiliations.

Copyright: © 2021 by the authors. Licensee MDPI, Basel, Switzerland. This article is an open access article distributed under the terms and conditions of the Creative Commons Attribution (CC BY) license (https://creativecommons.org/licenses/by/4.0/).

1. Introduction

The current agricultural system is based on the application of large quantities of chemical fertilizers that supply around 50% of the necessary nutrients for plant development [1]. However, the extensive use of this type of fertilization can lead to a wide range of environmental problems such as soil deterioration, the imbalance of nutrient proportions or water pollution. Therefore, the application of more eco-friendly fertilizers replacing the chemical ones has been considered one of the main agricultural challenges in recent years.

For several decades, many scientific projects have focused on plant growth promoting rhizobacteria (PGPR), one group of microorganisms that cannot only colonize the rhizosphere of crops but also improve their development through direct and indirect mechanisms [2]. Apart from enhancing plant development, PGPR bacteria are also nowadays being described as increasing food quality and improving the plant mineral content and bioactive compound concentrations [3], highly reported for its beneficial effects in human health [4].

As diets rich in fruit and vegetables are associated with a decrease in cardiovascular diseases and cancer, vegetables such as escarole are consumed in increasing amounts due to the beneficial effects for human health [5]. Escarole (*Cichorium endivia* L.) is attributed to contain antioxidant compounds, vitamins and phenolic compounds, which have been described as having a significant antioxidant activity [6] and therefore being one of the most economically important vegetables. Phenolic compounds are secondary metabolites that are synthesized in plants through the shikimic acid and phenylpropanoid pathways. They can be divided into two main families, flavonoids (which show a C6–C3–C6 general structural backbone) and non-flavonoids, among which phenolic acids are one of the most important groups in vegetables [7]. The direct scavenging of free radicals and reactive oxygen species (ROS) that can be attributed to these compounds is the main mechanism described for explaining their health protecting effects [8]. Moreover, these compounds are also important for food quality because they are closely related to food sensory characters such as bitterness, astringency and herbaceous flavors, thus affecting consumer acceptance [9].

Even though every year there are more published studies that increase the knowledge of PGPR bacteria effects in food quality, these improvements are still scarcely reported under abiotic stress conditions such as soil salinity, one of the major non-biological stresses [10]. Soil salinity severely limits plant development and affects agricultural productivity causing significant biochemical and physiological changes in crops. It affects plant growth and seed germination rates and induces water and oxidative stresses and nutritional disorders [11].

Scientific estimations calculate that soil salinity affects more than 800 million hectares of land worldwide and it is significantly increasing every year. Problems resulting from the augmentation in soil salinity are expanding worldwide because of global climate change, which has a great impact on the agricultural economy that has been growing for decades [10].

On the other hand, bacteria applied to fields as biofertilizers must be innocuous for human and environmental health and satisfy all food safety requirements [12], particularly on vegetables that are consumed raw such as escarole.

It must be remarked that bacterial strain effects vary from laboratories to the field and between different plant species due to an important host specificity. Some PGPR strains show a high potential to promote the growth of a particular crop and have no response in others [13]. *Bacillus halotolerans* and *Rhizobium laguerreae* are two bacterial species that have received the most extensive attention due to in vitro mechanisms and physiological traits and they have been previously reported for their beneficial plant effects in other crops of agri-food interest [11,14]. Related to the *Bacillus* genus, there are a few transcriptomic analyses that describe the genes involved in plant growth promotion and in biofilm formation [15] as well as in the production of secondary metabolites such as phenolic compounds or osmolytes [11]. Within the variety of species of the *Rhizobium* genus, the species *R. laguerreae* has been widely studied. We found several studies analyzing its interaction with legume plants. These strains formed biofilms and produced acyl-homoserine lactones (AHLs) involved in the quorum sensing regulation process [16]. They also solubilize phosphate, produce siderophores and symbiotic atmospheric nitrogen fixation [16]. Moreover, Ayuso-Calles and collaborators [17] reported a production of enzymes related to plant food quality such as derivatives of caffeoyl acid and quercetin. The aim of this study was to analyze for the first time the effects on phenolic composition, mineral content and plant growth development of escarole plants after the inoculation of probiotic bacterial strains from *Bacillus halotolerans* and *Rhizobium laguerreae* species under saline conditions.

2. Materials and Methods

2.1. Bacterial Strains

The bacterial strains used in this work were previously isolated from nodules of *Phaseolus vulgaris*. They were classified within the species *Rhizobium laguerreae* PEPV40 [14] and *Bacillus halotolerans* SCCPVE07 [11] by 16S rRNA gene analysis.

In this study, the PEPV40-GFP (green fluorescence protein), which contains the plasmid phC60, was used in fluorescence microscopy experiments. The PEPV40-GFP strain was also obtained in Jiménez-Gómez et al. [14].

PEPV40, PEPV40-GFP and SCCPVE07 were routinely grown at 28 °C in YMA (yeast manitol agar), TY (tryptone yeast, supplemented with 10 µg/mL tetracycline) and TSA (tryptic soy agar), respectively, for five days for *Rhizobium* strains and one day for *Bacillus*.

2.2. Draft Bacterial Genome Sequencing and Annotation

Bacterial genomic DNA was obtained from pure colonies of the PEPV40 strain grown on YMA plates and collected after 24 h at 28 °C using the ZR Fungal/Bacterial DNA MiniPrep (Zymo Research). The draft genome sequence of the bacterial isolates was obtained by shotgun sequencing on an Illumina MiSeq platform via a paired-end run (2×251 bp). The sequence data was assembled by Velvet 1.2.10 and gene annotation was performed using RAST 2.0 (Rapid Annotation using Subsystem Technology) [18].

The draft genome sequence of the strain *Rhizobium laguerreae* PEPV40 was deposited in GenBank under the accession number JABWPR000000000. The draft genome sequence of the strain *Bacillus halotolerans* SCCPVE07 was obtained previously in Jiménez-Gómez et al. [11].

2.3. Bacterial Colonization of Escarole Roots

Escarole seeds (*Cichorium endivia* L.) were surfaced-sterilized with a solution of $HgCl_2$ 0.1% for four minutes. This type of sterilization was also used for in vitro and greenhouse experiments. The seeds were then washed five times with sterile water and germinated on water-agar plates. Seedlings were kept in the dark at 28 °C for two days and then transferred to 1.5% agar square plates (12×12 cm) with a distribution of six seedlings per plate.

According to the bacterial strain applied, the colonization assays consisted of two different treatments. On the one hand, escarole roots were inoculated with PEPV40-GFP and observed directly at seven and fourteen days post inoculation (dpi). On the other hand, escarole roots inoculated with the SCCPVE07 strain were immersed in an Uricase antibody preparation following the instructions described in Jiménez-Gómez et al. [11] before microscopy analysis, which was also carried out at seven and fourteen dpi.

To prepare the PEPV40-GFP bacterial suspension, the strain was grown for five days at 28 °C on $TYTc^{10}$ plates. The SCCPVE07 strain was grown for one day at 28 °C on TSA plates. Both suspensions were adjusted to an OD (optical density) (600 nm) of 0.6, which corresponded with a final concentration of 10^8 CFU/mL after counting the number of viable cells using the serial decimal dilution method.

Propidium iodide of 10 µm was added into the roots of both treatments 10 min before microscopy root observations. Fluorescence microscopy was carried out with a Nikon Eclipse 80i fluorescence microscope and a mercury lamp was used for green fluorescent protein excitation.

2.4. Effects of Bacterial Strains on Plant Seedlings

Surfaced-sterilized escarole seedlings were transferred to 0.2 dm^3 pots that contained sterilized vermiculite "SEED PRO 6040" (PROJAR, Madrid, Spain) as a substrate. The escarole plants were inoculated with 2 mL of bacterial suspension at 10^8 CFU mL^{-1} (0.6 OD measured at 600 nm). Fifty-four pots were used as a negative control and fifty-four pots were added per each bacterial suspension. In each treatment half of the plants were irrigated with water from a bottom reservoir every 48 h and the other half with a saline condition (NaCl 100 mM). The pots were maintained in a growth chamber. Fifteen days post inoculation, the data of fresh weight per plant, dry weight per plant and leaf size were recorded. The experiment was performed at least three times. Therefore, 81 plants per treatment and growth condition were collected for obtaining the data.

2.5. Evaluation of Escarole Growth Promotion and Nutritional Content

The ability to promote plant growth was investigated on escarole plants using a mix of 2.4 L of non-sterilized soil and vermiculite (3:1 v/v) as a substrate in plastic pots.

The escarole seeds were surface-sterilized and pre-germinated. They were transferred into the substrate and inoculated with 5 mL of bacterial suspension with a final concentration of 10^8 CFU mL^{-1} (0.6 OD measured at 600 nm). Uninoculated escarole plants were included as negative controls under the same conditions. The plants were watered from a bottom reservoir every 48–72 h depending on plant demand. The solution of NaCl 100 mM (to obtain saline conditions) was added in the specific treatments. The plants were maintained in an illuminated greenhouse (night temperatures ranging from 15 to 20 °C and day temperatures ranging from 25 to 30 °C) with a humidity control for 60 days. A total of 18 plants were used in each treatment. The values of the number of leaves, shoot fresh weight and shoot dry weight per plant were recorded.

The dry plants were used for the analysis of N, C, P, Mg, K and Fe contents. The analysis was performed by the Ionomics Service at CEBAS-CSIC (Murcia, Spain), using elemental analyst model TruSpec CN628 equipment for the N analysis and ICP THERMO ICAP 6500DUO equipment for the analysis of the remaining elements.

2.6. Evaluation and Analysis of Phenolic Compound Contents

After 60 days post bacterial inoculation, the phenolic composition of the plant material was determined as previously reported [11]. Freeze-dried samples (5 mg) were extracted in a bath of ultrasound using MeOH:H$_2$O (80:20). An extraction was performed three times and supernatants were combined and cleaned by liquid-liquid extraction with hexane. Organic solvents were removed under reduced pressure until a final volume of 2 mL was reached.

HPLC-DAD-MS analyses were performed to determine the phenolic composition of the extracts by using a Hewlett–Packard 1200 series liquid chromatograph (Agilent Technologies, Waldbronn, Germany). Chromatography was carried out in a Spherisorb®S3 ODS-2 C18 reversed phase 3 μm 150 × 4.6 mm column (Waters Corporation, Milford, MA, USA) thermostatted at 35 °C using formic acid (0.1 mL L^{-1}) and acetonitrile (B) as solvents [11]. The preferred wavelengths employed for detection were 280, 330 and 370 nm and full UV-vis spectra were recorded from 220 to 600 nm. Mass spectrometry was carried out using an API 3200 Qtrap (Applied Biosystems, Darmstadt, Germany) equipped with an Electrospray Ionization (ESI) source and a triple quadrupole-ion trap mass analyzer and employed a previously reported methodology [11]. Spectra were recorded in negative ion mode between m/z 100 and 1700 and both full scan and MS/MS analyses were performed.

The chromatographic retention times, UV-vis spectra and mass spectra (m/z and fragmentation patterns) were used for compound identification. The quantification was performed by an external standard method from the peak area values obtained in the chromatograms recorded at 280 nm (protocatechuic acid derivative), 330 nm (caffeic acid and its derivatives including cichoric acid) or 360 nm (flavone and flavonol derivatives). The content of protocatechuic acid and caffeic acid derivatives were expressed as gallic acid and caffeic acid equivalents, respectively. Kaempferol and quercetin derivatives were expressed as kaempferol 3-O-glucuoside and quercetin 3-O-glucoside equivalents, respectively. The extraction and analysis were performed in triplicate and the results provided are the mean values expressed in g Kg^{-1} of plant dry weight.

2.7. Statistical Analysis

The statistical analysis was performed using the software program StatView 5.0 (SAS Institute Inc., Cary, NC, USA). An analysis of variance (ANOVA) followed by Fisher's protected least significant difference test (Fisher's PLSD) were performed as the statistical analysis.

3. Results and Discussion

3.1. Bacterial Genome Mining

Nowadays, bacterial genome mining strategies and studies are considered a useful tool to identify functionally bacterial properties that provide further insights into plant-microbe interactions and new biofertilizer formulations [19].

Although many mechanisms have been described to explain plant growth promotion by bacteria, a single mechanism is not responsible for the full positive effects. Several mechanisms rather than just one participate in the beneficial association [20].

As described in Jiménez-Gómez et al. [11], the SCCPVE07 strain solubilizes phosphate and produces siderophores, exopolysaccharides and phytohormones. Moreover, a wide variety of enzymes involved in flavonoid and phenolic acid metabolisms were reported. In this study, it was observed that apart from the previously published work the SCCPV07 bacterial genome encoded the gene involved in aromatic amino acid aminotransferase (AAAAT) (EC 2.6.1.57) biosynthesis, which is a precursor of phenylalanine. Moreover, the gene that encodes the catechol-2,3-dioxygenase enzyme (EC 1.13.11.2) presented in the protocatechuic acid pathway was also reported. Both are interesting in the active compounds in plants [21].

Among the range of essential plant nutrients, phosphorus is a good example required in plant growth and development. Currently, phosphorus deficiency is one of the major issues and limitations to crop production. Furthermore, it is estimated that the phosphorus consumption in the agricultural system is increasing around 2.5% per year [22]. In this sense, the use of PGPR bacteria in the increase of phosphorus efficiency through making insoluble phosphorus compounds available for plants is currently a focus of special attention [11]. The *R. laguerreae* PEPV40 encodes several enzymes involved in the transformation of inorganic substrates into more easily assimilable chemical forms for plants such as pyrophosphatase (EC 3.6.1.1) or citrate synthase (EC 2.3.3.1).

Microorganisms have evolved to produce siderophores, which are secreted to solubilize iron from their surrounding areas. Apart from to satisfy nutritional requirements, it is also a key point to cope with other bacteria in the plant rhizosphere. Phytopathogenic bacteria or fungi can be deprived of iron availability and therefore the production of antibiotics is extremely decreased [23]. The PEPV40 genome was observed to encode genes for iron acquisition and siderophore biosynthesis such as aerobactin. The biosynthesis above described led to a notable increase in iron and phosphorus plant content in the escarole leaves analyzed (Table 1).

Table 1. Effects of *Rhizobium laguerreae* PEPV40 and *Bacillus halotolerans* SCCPVE07 inoculation on the nutrient contents of escarole plants.

	N (g/100 g)				C (g/100 g)			
	Non-Saline Conditions		100 mM NaCl		Non-Saline Conditions		100 mM NaCl	
	Mean ± SE	%	Mean ± SE	%	Mean ± SE	%	Mean ± SE	%
Control	3.69 ± 0.10		2.90 ± 0.05		41.88 ± 0.15		35.74 ± 0.47	
PEPV40	3.58 ± 0.01	-	3.36 ± 0.01 *	+15.8	43.25 ± 0.27	+3.3	39.07 ± 0.23*	+9.3
SCCPVE07	3.24 ± 0.22	-	2.74 ± 0.05	-	41.53 ± 1.20	-	35.18 ± 0.32	-

	P (g/100 g)				Mg (g/100 g)			
	Non-Saline Conditions		100 mM NaCl		Non-Saline Conditions		100 mM NaCl	
	Mean ± SE	%	Mean ± SE	%	Mean ± SE	%	Mean ± SE	%
Control	0.60 ± 0.02		0.40 ± 0.01		0.67 ± 0.03		0.46 ± 0.01	
PEPV40	0.88 ± 0.03 *	+46.7	0.57 ± 0.01 *	+42.5	0.90 ± 0.07 *	+34.3	0.66 ± 0.01 *	+43.5
SCCPVE07	0.75 ± 0.01 *	+25.0	0.35 ± 0.02	-	1.04 ± 0.01 *	+55.2	0.33 ± 0.02	-

	K (g/100 g)				Fe (mg/Kg)			
	Non-Saline Conditions		100 mM NaCl		Non-Saline Conditions		100 mM NaCl	
	Mean ± SE	%	Mean ± SE	%	Mean ± SE	%	Mean ± SE	%
Control	2.90 ± 0.08		3.18 ± 0.07		0.51 ± 0.04		0.10 ± 0.01	
PEPV40	1.83 ± 0.09	-	2.50 ± 0.07	-	0.69 ± 0.06 *	+35.3	0.94 ± 0.03 *	+840
SCCPVE07	2.87 ± 0.04	-	3.61 ± 0.22 *	+13.5	1.14 ± 0.02 *	+123.5	0.12 ± 0.01	+20.0

Table 1. Cont.

	Na (g/100 g)			
	Non-Saline Conditions		100 mM NaCl	
	Mean ± SE	%	Mean ± SE	%
Control	0.51 ± 0.02		4.07 ± 0.12	
PEPV40	0.30 ± 0.01 *	-	2.57 ± 0.07 *	-
SCCPVE07	0.52 ± 0.06	+1.96	4.57 ± 0.49 *	+12.3

* indicates a significant difference between inoculated values and the negative control, $p \leq 0.05$ according to Fisher's Protected LSD (least significant differences). SE = standard error.

According to the importance of the colonization process for a successful plant-microbe interaction, the mechanisms involved in bacterial root attachment were analyzed. The PEPV40 bacterial genome encodes genes involved in cellulose production [beta-1,4-glucanase (cellulase) (EC 3.2.1.4)], broadly described as essential in the colonization process.

The functional analysis of the production of phytohormones has been thoroughly investigated due not only to the agronomic impact on horticultural crops but also to the significant improvement in nutrient absorption. However, plants are not the only organisms with the ability to produce them.

Indol-3-acetic acid (IAA) is reported as one of the most influential phytohormones in enhancing plant growth promotion [20]. In this sense, the PEPV40 strain genome encodes genes related to the biosynthesis of indol-3-acetic acid (Indole-3-glycerol phosphate synthase (EC 4.1.1.48) and it also encodes AmiE acetamidases (EC 3.5.1.4) involved in the synthesis of IAA by the so-called indole-3-acetamide pathway by converting indole-3-acetamide to indole acetic acid.

As mentioned before, soil salinity is a worrying issue in agriculture today. Previous research studies have discovered that the plant accumulations of solutes such as choline or proline act as a natural defense system from salinity. In this sense, the application of microorganisms with the ability of this solutes production can be an effective and hopeful solution. The PEPV40 genome encodes *betA* and *betB* genes, which are involved in the codification of choline dehydrogenase (EC 1.1.99.1) and betaine aldehyde dehydrogenase (BADH), respectively. As it can be observed in Table 1, under salinity stress conditions the inoculation with the strain PEPV40 did not increase the sodium content and the results obtained in the control treatment and the plants inoculated with SCCPVE07 strain were similar.

As phenolic compounds have been related to both health protective effects and food quality [8,9], bacterial genes potentially involved in the biosynthesis or transport of phenolic compounds were also searched for in the PEPV40 genome. The PEPV40 genome shows a wide variety of important genes in flavonoid and phenolic acid metabolisms.

We searched for genes involved in the caffeic acid pathway, which is involved in lignin synthesis and interacts with reactive oxygen species. The biosynthesis of ferulic acid is carried out through caffeic acid whose process is mediated by the enzyme 3-O-methyltransferase [EC 2.1.1.68], which was not detected in the in silico analysis. However, several genes involved in previous steps in the synthesis route of phenylpropanoid compounds were detected such as the gene that encodes the enzyme cinnamoyl esterase. Furthermore, the gene encoding the enzyme trans-hydratase-feruloyl-CoA, involved in the production of vanillic acid from ferulic acid as a substrate, was also detected.

The presence of the enzyme's aromatic amino acid aminotransferase (AAAAT) (EC 2.6.1.57) and EC 2.3.1.74 and a naringenin-chalcone synthase were also observed. The third one is involved in obtaining naringenin from phenylalanine, a flavone that acts as a precursor in the biosynthesis pathways of kaempferol, quercetin and apigenin [24].

Multiple bacterial strains of *Rhizobium* and *Bacillus* genera have also been previously described as containing a wide variety of genes involved in promoting plant growth in their genomes. Ambreetha and collaborators [25] reported a significant modification in the architecture of rice roots modulated by the expression of *Bacillus* auxin-responsive genes.

On the other hand, Pérez-Montaño et al. [26] reported many *Rhizobium* genes that could represent a strategy to establish symbiosis under salt stressing conditions.

Based on the study described in Jiménez-Gómez et al. [11] and the results here presented, strains SCCPVE07 and PEPV40 exhibited an interesting potential as plant growth promoting microorganisms. However, these abilities were also checked and presented in planta even under salinity stress conditions with the following results.

3.2. Bacterial Colonization Analysis of Escarole Roots

It was observed that the PEPV40 strain colonized escarole root surfaces, forming microcolonies of biofilm initiation in the intercellular spaces in the cortical cells on both days of analysis (Figure 1 A–C). On the other hand, Figure 1 reveals that the SCCPVE07 strain (Figure 1 B–D) also colonized root surfaces and this was significantly so in the base of emergent lateral roots. Both strains increased the colonization gradually during the observations.

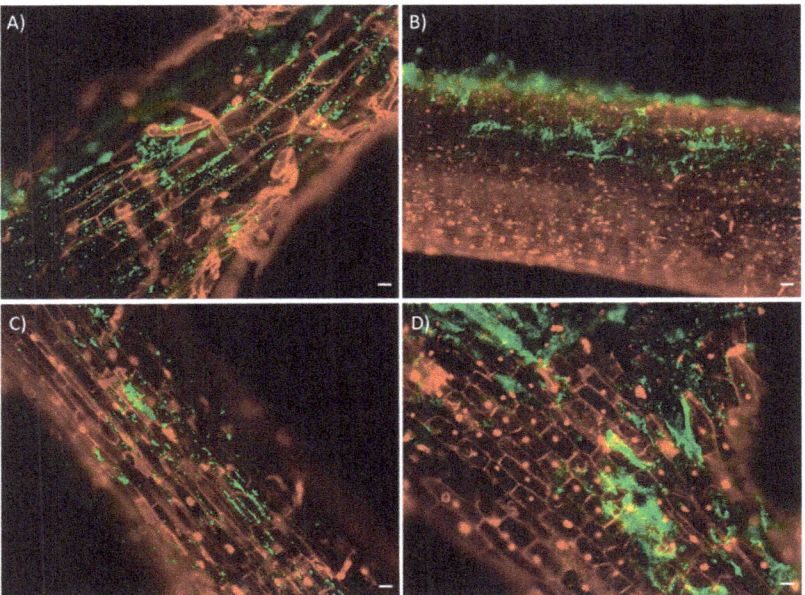

Figure 1. Bacterial colonization of escarole roots. Fluorescence optical micrographs of escarole seedlings roots inoculated with the PEPV40 strain (**A,C**) and SCCPVE07 (**B,D**) obtained at 7 (**A,B**) and 14 (**C,D**) days after inoculation. The micrographs show the bacterial ability (green color) to colonize the root surfaces, the base of emerging roots and the initiation of microcolonies. White bar of **A**, **C** and **D** images, 100 µM. Bar of **B** image, 200 µM.

Bacterial root colonization is a key point required for PGPR bacteria. The adhesion is related to the production of a wide range of metabolites involved in plant-microbe interaction or to the ability to mobilize plant nutrients. Moreover, it is important in studies of bacterial persistence and understanding when microbes are applied in the soil [27]. In this sense, biofilm formation is recognized for its beneficial role in the PGP traits and the improvement in bacterial persistence in plant roots [28].

The colonization dynamics here reported have been previously described in other PGPR strains of the same genera. Our results showed that PEPV40 and SCCPVE07 not only presented the same distribution as other bacterial PGPR strains from *Rhizobium* and *Bacillus* genera, respectively [29,30], but also their colonization patterns were similar to other strains in different horticultural crop roots [11,27]. However, this study is the first to report *Rhizobium laguerreae* and *Bacillus halotolerans* strains as efficient colonizers of escarole roots.

3.3. Effects of Bacterial Strains on Escarole Growth and Nutritional Content

The results of escarole growth promotion in the in vitro and greenhouse experiments (Table 2) showed that PEPV40 and SCCPVE07 strains improved the growth parameters evaluated in non-stress and under salt stress conditions.

Table 2. Results from in vitro growth promotion experiments and effects of *Rhizobium laguerreae* PEPV40 and *Bacillus halotolerans* SCCPVE07 inoculation on the growth and number of leaves of escarole plants under greenhouse conditions.

	Results from In Vitro Growth Promotion Experiments							
	Stem Length (cm)				SFW(g)			
	Non-Saline Conditions		100 mM NaCl		Non-Saline Conditions		100 mM NaCl	
	Mean ± SE	%	Mean ± SE	%	Mean ± SE	%	Mean ± SE	%
Control	14.0 ± 0.3		16.0 ± 0.2		0.33 ± 0.02		0.36 ± 0.01	
PEPV40	16.0 ± 1.0 *	+14.3	18.0 ± 3.4 *	+12.5	0.39 ± 0.03 *	+18.2	0.40 ± 0.05	+13.9
SCCPVE07	15.0 ± 0.3	+7.1	18.0 ± 0.2	+12.5	0.36 ± 0.01	+9.1	0.45 ± 0.01 *	+25.0
	Results from In Vitro Greenhouse Experiments							
	SDW(g)				SFW(g)			
	Non-Saline Conditions		100 mM NaCl		Non-Saline Conditions		100 mM NaCl	
	Mean ± SE	%	Mean ± SE	%	Mean ± SE	%	Mean ± SE	%
Control	42.18 ± 0.55		27.07 ± 0.67		1.76 ± 0.02		1.13 ± 0.01	
PEPV40	49.81 ± 0.53 *	+18.1	31.18 ± 0.61 *	+15.2	2.07 ± 0.02 *	+17.6	1.30 ± 0.03 *	+15.0
SCCPVE07	43.64 ± 0.61	+3.5	47.63 ± 0.62 *	+76.0	1.82 ± 0.03	+3.4	1.99 ± 0.03 *	+76.1
	Results from In Vitro Growth Promotion Experiments				Results from In Vitro Greenhouse Experiments			
	Stem Length (cm)				Number of Leaves			
	Non-Saline Conditions		100 mM NaCl		Non-Saline Conditions		100 mM NaCl	
	Mean ± SE	%	Mean ± SE	%	Mean ± SE	%	Mean ± SE	%
Control	4.53 ± 0.27		4.69 ± 0.16		12.87 ± 0.44		12.60 ± 0.43	
PEPV40	5.43 ± 0.15	+19.9	4.71 ± 0.11	+0.4	14.67 ± 0.47 *	+14.0	12.47 ± 0.39	-
SCCPVE07	4.61 ± 0.21	+1.8	5.34 ± 0.11 *	+13.9	14.47 ± 0.61 *	+12.4	13.20 ± 0.39	+4.8

* indicates a significant difference between inoculated values and the negative control, $p \leq 0.05$ according to Fisher's Protected LSD (least significant differences). SE = standard error. SFW = shoot fresh weight. SDW = shoot dry weight.

On the one hand, in the in vitro experiments the highest values were always shown in escarole plants inoculated with a bacterial treatment. In non-stress conditions, the stem length was increased by 19.9% and the shoot dry weight up to 14.3% in the case of the PEPV40 strain application compared with the respective control. However, under salt stress the best values were reported after the application of the SCCPVE07 strain. In particular, the largest increase (25.0%) was shown in the shoot fresh weight.

On the other hand, the results from the greenhouse experiments also showed that in a general way the best values were described in escarole plants after bacterial strain applications. In non-stress conditions, the highest growth values were always reported in plants inoculated with the PEPV40 strain. The number of leaves was significantly increased by 14.0% and the shoot fresh and dry weight were 18.1% and 17.6%, respectively, compared with the control. In the same way, under salt stress conditions and the SCCPVE07 strain, the escarole plants showed the best shoot fresh (76.0%) and dry (76.1%) values regarding the respective control treatment.

The results of the present study showed important improvements in nutrient plant content compared with their respective plants in the un-inoculated treatments.

The enhancement in the specific nutrient content was probably related to those genes encoded in bacterial genomes, previously described for their relation in the ability to mobilize plant nutrients. Most values were statistically higher after the bacterial application. However, the phosphorus, magnesium and iron showed the greatest improvements. The magnesium content showed up to 34.3% in non-saline conditions and 43.5% under saline stress both after PEPV40 application and compared with their respective controls. The plant content of elements such as nitrogen required for crops was also significantly increased after the PEPV40 bacterial application; up to 15.8% under saline stress conditions.

The plant assays carried out in a chamber and the greenhouse broadly showed that *B. halotolerans* and *R. laguerreae* promoted the plant growth of escarole crops in both growth conditions. In the initial stages of growth and in vitro experiments, the escarole seedlings inoculated had longer stems than the plants of uninoculated controls, suggesting that both strains exerted a beneficial effect on plant growth and development. Moreover, in the greenhouse experiments the number of leaves was also higher, an interesting point in escarole cultivation.

Other studies with bacterial strains from the same genera have also reported significant improvements. A similar increase in leaf mass and grain yield but in fava bean plants was also detected after the application with a biofertilizer based on a *R. laguerreae* strain [31]. After the application of a *B. halotolerans* strain, El-Akhdar and collaborators [32] reported an enhancement not only in wheat plant promotion but also in nitrogen plant content.

Finally, the results here reported indicated for the first time an improvement in the nutrient content and escarole growth after the application of a *Bacillus* and *Rhizobium* bacterial strain. Moreover, as far as we know, this was the first study performed with escarole (*Cichorium endivia* L.) crops.

3.4. Effects of Bacterial Strains on Plant Phenolic Composition

The analysis performed allowed for the determination of 15 different phenolic compounds in escarole leaves. To be precise, 10 phenolic acids (one hydroxybenzoic acid and nine hydroxycinnamic acid derivatives) and five flavonols (one quercetin derivative and four kaempferol derivatives) were identified and quantified in the samples (Table 3). Phenolic acids accounted around 70% of total phenolic compounds acids with cichoric acid the most abundant followed by caffeoyl-quinic acid, caffeoyl-tartaric acid and protocatechuic acid glucoside (Table 4). With regard to flavonols, this family of phenolic compounds accounted for around 30% of total phenolic compounds with kaempferol derivatives the most abundant compound and in particular the glucuronide derivative (Table 5). This phenolic composition agreed with those previous reported [5,9,17] although some compounds such as the protocatechuic acid derivative were identified in this work for the first time.

Table 3. Chromatographic, spectral and spectrometric features of the phenolic compounds identified in escarole plants.

Compound	t_R (min)	UV λ_{max} (nm)	[M−H] (m/z)	Fragment Ions (m/z)	Identification
1	10.58	284	315	153	Protocatechuic acid glucoside
2	11.09	332–295 (sh)	311	179,135	Caffeoyl-tartaric acid
3	12.24	338–292 (sh)	343	179,135	Caffeic acid derivative
4	14.01	328–300 (sh)	353	191	Caffeoyl-quinic acid
5	15.24	328–300 (sh)	353	191	Caffeoyl-quinic acid
6	16.8	324–296 (sh)	179	135	Caffeic acid
7	17.3	328–300 (sh)	295	179,135	Caffeoyl-malic acid
8	21.37	328–300 (sh)	473	311,293,179,149,135	Cichoric acid
9	27.9	-	477	301	Quercetin 3-O-glucuronide
10	29.34	328–300 (sh)	487	307,293,193,179,135	Caffeoyl-ferouyl-tartaric acid
11	29.9	-	572	461,397,285	Kaempferol glucuronide derivative
12	30.78	328–300 (sh)	515	353,191,179,135	Dicaffeoyl-quinic acid
13	32.6	348–296 (sh)	461	285	Kaempferol 3-O-glucuronide
14	33.06	348–296 (sh)	447	285	Kaempferol 3-O-glucoside
15	36.00	348–296 (sh)	533	489, 285	Kaempferol 3-O-malonyl-glucoside

Table 4. Concentration, expressed as mean ± standard deviation, of phenolic acids of escarole plants.

	Non-Saline Conditions (g Kg^{-1})			100 mM NaCl (g Kg^{-1})		
	Control	PEPV40	SCCPVE07	Control	PEPV40	SCCPVE07
Protocatechuic acid glucoside	8.39 ± 0.06 c	13.28 ± 0.06 b	14.0 ± 0.1 a	3.8 ± 0.2 b, *	6.42 ± 0.07 a, *	3.11 ± 0.06 c, *
Caffeoyl-tartaric acid	6.10 ± 0.03 a	4.51 ± 0.01 c	5.85 ± 0.06 b	4.9 ± 0.3 b, *	6.45 ± 0.08 a, *	3.94 ± 0.02 c, *
Caffeic acid derivative	0.43 ± 0.02 a	0.43 ± 0.02 a	0.43 ± 0.03 a	0.41 ± 0.02 b	0.53 ± 0.01 a, *	0.381 ± 0.003 b
Caffeoyl-quinic acids [1]	13.21 ± 0.05 a	10.40 ± 0.07 b	8.58 ± 0.02 c	8.9 ± 0.5 b, *	12.97 ± 0.08 a, *	9.325 ± 0.008 b, *
Caffeic acid	0.64 ± 0.01 a	0.53 ± 0.03 c	0.582 ± 0.003 b	0.48 ± 0.02 b, *	0.67 ± 0.006 a, *	0.436 ± 0.009 c, *
Caffeoyl-malic acid	0.60 ± 0.01 a	0.46 ± 0.02 b	0.45 ± 0.03 b	0.41 ± 0.01 b, *	0.59 ± 0.01 a, *	0.37 ± 0.01 c, *
Cichoric acid	37.1 ± 0.2 a	24.8 ± 0.2 c	25.76 ± 0.09 b	24.56 ± 1.35 b, *	34.51 ± 0.04 a, *	23.38 ± 0.07 b, *
Caffeoyl-ferouyl-tartaric acid	1.21 ± 0.03 a	0.47 ± 0.02 c	1.03 ± 0.04 b	1.51 ± 0.07 b, *	2.50 ± 0.07 a, *	0.82 ± 0.03 c, *
Dicaffeoyl-quinic acid	1.37 ± 0.06 a	0.81 ± 0.03 b	1.42 ± 0.02 a	1.35 ± 0.06 a	1.23 ± 0.04 b, *	0.94 ± 0.02 c, *
Total phenolic acids	69.0 ± 0.1 a	55.7 ± 0.5 c	58.1 ± 0.3 b	46 ± 2 b, *	65.87 ± 0.07 a, *	42.71 ± 0.08 c, *

[1] Sum of caffeoyl quinic acids. Different letters within each type of conditions (saline or non-saline) and each row indicates Scheme 0.
* indicates significant differences ($p \leq 0.05$ according to Fisher's Protected LSD) between the saline sample and the corresponding non-saline sample.

Table 5. Flavonol composition of escarole plants, expressed as mean ± standard deviation.

	Non-Saline Conditions (mg/g)			100 mM NaCl (mg/g)		
	Control	PEPV40	SCCPVE07	Control	PEPV40	SCCPVE07
Kaempferol 3-O-glucoside	2.20 ± 0.08 b	2.18 ± 0.01 b	2.64 ± 0.05 a	2.12 ± 0.11 b	2.64 ± 0.05 a, *	2.16 ± 0.07 b, *
Kaempferol 3-O-glucuronide	15.58 ± 0.10 b	15.30 ± 0.09 b	18.86 ± 0.06 a	13.60 ± 0.45 b	19.76 ± 0.06 a	14.12 ± 0.04 b, *
Kaempferol-glucuronide derivative	1.19 ± 0.03 a	0.77 ± 0.03 b	0.80 ± 0.02 b	0.97 ± 0.07 b, *	1.45 ± 0.06 a, *	1.00 ± 0.04 b, *
Kaempferol 3-O-malonyl-glucoside	4.95 ± 0.08 c	5.42 ± 0.08 b	5.91 ± 0.09 a	4.1 ± 0.3 b, *	5.11 ± 0.06 a, *	4.90 ± 0.09 a, *
Quercetin 3-O-glucuronide	0.31 ± 0.01 a	0.23 ± 0.02 b	0.29 ± 0.01 a	0.21 ± 0.02 b, *	0.303 ± 0.009 a, *	0.23 ± 0.01 b, *
Total flavonoids	24.2 ± 0.4 b	23.9 ± 0.3 b	28.5 ± 0.1 a	21 ± 1 b, *	29.3 ± 0.1 a, *	22.4 ± 0.3 b, *

Different letters within each type of conditions (saline or non-saline) and each row indicates significant differences ($p \leq 0.05$). * indicates significant differences ($p \leq 0.05$ according to Fisher's Protected LSD) between the saline sample and the corresponding non-saline sample.

The salinity level of grown media had an important effect on the phenolic content of the escarole leaves of non-inoculated plants. In the case of both phenolic acids and flavonols (Tables 4 and 5), the content of all of the detected compounds and therefore the total content was significantly lower when plants were grown under saline conditions than under non-saline conditions. This behavior was different from that observed in other plants such as lettuce [33] but it has been reported that the effect of salinity on the phenolic content of herbaceous plants was strongly dependent on both plant variety and the type of salinity to which it was exposed [34]. These results point out for the first time that the biosynthesis and/or accumulation of phenolic compounds in the leaves of *Cichorium endivia* L. plants was negatively affected. Thus, the different strategies that can ease those negative effects such as the use of PGPR bacteria proposed in this work need to be assessed.

The results obtained pointed out that the effect of bacterial inoculation on the phenolic composition of escarole leaves was different depending not only on the phenolic compound but also on the growth conditions and the bacterial strain employed. Under non-saline conditions, the inoculation with PEPV40 or SCCPVE07 strains led to lower levels of total phenolic acids (Table 4). Although the content of some phenolic compounds was not or only slightly affected by inoculation, when compared with the non-inoculated plants, the inoculated plants showed significantly lower levels of the most abundant phenolic compounds detected, i.e., cichoric and caffeoyl-quinic acids. On the contrary, the content of some phenolic acids such as the protocatechuic acid derivative was significantly higher in the inoculated plants, mainly when the SCCPVE07 strain was used. In fact, it was observed that the SCCPV07 bacterial genome included genes encoding enzymes involved in the protocatechuic acid pathway, which may explain the higher concentration. As for the flavonoid content (Table 5), the inoculation with SCCPV07 in plants grown under non-saline conditions led to significantly higher levels of flavonols in the escarole leaves. It seems that the inoculation with this strain favored the biosynthesis of flavonols in escarole leaves, mainly of the most abundant kaempferol derivatives (glucuronide and malonyl-glucoside derivatives). The inoculation with PEPV40 under these conditions did not significantly modify the total flavonol content of the escarole leaves although it led to significantly higher levels of kaempferol 3-O-malonyl glucoside, the second most important flavonol detected in the plants.

When the plants were grown under saline conditions, the inoculation with the PEPV40 strain significantly increased the total concentration of both phenolic acid and flavonols in the escarole leaves (Tables 4 and 5). As aforementioned, the genome of this strain showed a wide variety of important genes involved in the biosynthesis pathway of flavonols and phenolics acids, which might explain those contents. In fact, the inoculation with this strain not only alleviated the negative effects that salinity had on the phenolic content but also, in the case of flavonols, led to higher contents than the control plants grown under non-saline conditions (Table 5). Thus, the use of this type of PGPR bacteria under salinity stress grown conditions could improve the escarole quality with regard to its phenolic composition.

The effect of SSCPVE07 inoculation in the plants grown under saline conditions differed depending on the type of phenolic compounds. Hence, under those conditions, the plant inoculated with this strain showed lower levels of total phenolic acids. However, slightly higher levels of flavonols (although the differences were not significant) when compared with control samples were detected in the SSCPVE07 inoculated plants. These levels were similar to those found in the control samples grown under non-saline conditions, which pointed out that the inoculation with this strain could ease the negative effects that salinity stress exerted on the accumulation of flavonols in escarole leaves.

Hence, although both PGPR bacteria strains assayed could alleviate the negative effects of salinity stress on the phenolic composition of *Cichorium endivia* L. plants, the PEPV40 strain provided the best results, even improving the phenolic contents of escarole leaves from plants grown under non-stressful conditions.

4. Conclusions

This study is the first report on the effects of the inoculation of escarole plants with *R. laguerreae* a *B. halotolerans* strains. The results showed how bacterial inoculation led to significant increases not only in the plant development but also in the nutritional content and food quality even under saline stress conditions. The edible parts of escarole inoculated plants showed significant increases in the content of phenolics acids such as cichoric acid and caffeoyl-tartaric acid as well as flavonoids such as kaempferol 3-O-glucuronide. Moreover, a relevant enhancement in nitrogen, phosphorus and magnesium content was also observed. The results showed that inoculation with *R. laguerreae* PEPV40 and *B. halotolerans* SCCPVE07 have an interesting agronomic potential, improving the content of several phenolic compounds of escarole plants. Furthermore, the bases for this type of fertilization could be established in this crop even under conditions of saline stress, maintaining food safety and increasing plant development.

Author Contributions: Conceptualization, R.R. and P.G.-F.; methodology, A.J.-G. and I.G.-E; software, A.J.-G., I.G.-E and M.T.E.-B.; validation, R.R., P.G.-F, I.G.-E., M.T.E.-B. and A.J.-G.; formal analysis, A.J.-G., I.G.-E., M.T.E.-B. and R.R.; investigation, A.J.-G.; resources, R.R and M.T.E.-B; data curation, R.R., P.G.-F., I.G.-E., M.T.E.-B. and A.J.-G.; writing—original draft preparation, A.J.-G; writing—review and editing, R.R., P.G.-F., I.G.-E., M.T.E.-B. and A.J.-G.; visualization, R.R. and M.T.E.-B; supervision, P.G.-F., R.R and M.T.E.-B; project administration, A.J.-G. and R.R; funding acquisition, R.R and M.T.E.-B. All authors have read and agreed to the published version of the manuscript.

Funding: This work was supported by Grants AGL2015-70510-R from MINECO (Spanish Ministry of Economy, Industry and Competitiveness) and Strategic Research Programs for Units of Excellence from Junta de Castilla y León (CLU-2O18-04). AJG is the recipient of a FPU predoctoral fellowship from the Central Spanish Government. IGE thanks the Spanish MICINN (Ministry of Science, In-novation and Universities) for the Juan de la Cierva-incorporación postdoctoral contract (IJCI-2017-31499).

Institutional Review Board Statement: Not applicable.

Informed Consent Statement: Not applicable.

Data Availability Statement: Not applicable.

Acknowledgments: The authors thank Edurne Arroyo Torres for editing the English language of the manuscript submitted for publication.

Conflicts of Interest: The authors declare no conflict of interest.

References

1. Li, H.; Qiu, Y.; Yao, T.; Ma, Y.; Zhang, H.; Yang, X. Effects of PGPR microbial inoculants on the growth and soil properties of *Avena sativa, Medicago sativa,* and *Cucumis sativus* seedlings. *Soil Tillage Res.* **2020**, *199*, 104577. [CrossRef]
2. Kumar, B.D.; Jacob, J. Plant growth promoting rhizobacteria as a biological tool for augmenting productivity and controlling disease in agriculturally important crop-A review. *J. Spices Aromat. Crop.* **2019**, *28*, 77–95.
3. Pagnani, G.; Pellegrini, M.; Galieni, A.; D'Egidio, S.; Matteucci, F.; Ricci, A.; Del Gallo, M. Plant growth-promoting rhizobacteria (PGPR) in *Cannabis sativa* 'Finola'cultivation: An alternative fertilization strategy to improve plant growth and quality characteristics. *Ind. Crop. Prod.* **2018**, *123*, 75–83. [CrossRef]
4. Zielińska, M.A.; Białecka, A.; Pietruszka, B.; Hamułka, J. Vegetables and fruit, as a source of bioactive substances, and impact on memory and cognitive function of elderly. *Postępy Hig. I Med. Doświadczalnej* **2017**, *71*, 267–280. [CrossRef]
5. Llorach, R.; Martínez-Sánchez, A.; Tomás-Barberán, F.A.; Gil, M.I.; Ferreres, F. Characterisation of polyphenols and antioxidant properties of five lettuce varieties and escarole. *Food Chem.* **2008**, *108*, 1028–1038. [CrossRef] [PubMed]
6. Sinbad, O.O.; Folorunsho, A.A.; Olabisi, O.L.; Ayoola, O.A.; Temitope, E.J. Vitamins as Antioxidants. *J. Food Sci. Nutr. Res.* **2019**, *2*, 214–235. [CrossRef]
7. De La Rosa, L.A.; Moreno-Escamilla, J.O.; Rodrigo-García, J.; Alvarez-Parrilla, E. Phenolic Compounds. In *Postharvest Physiology and Biochemistry of Fruits and Vegetables*; Elsevier BV: Amsterdam, The Netherlands, 2019; pp. 253–271.
8. Shahidi, F.; Ambigaipalan, P. Phenolics and polyphenolics in foods, beverages and spices: Antioxidant activity and health effects-A review. *J. Funct. Foods* **2015**, *18*, 820–897. [CrossRef]
9. D'Antuono, L.F.; Ferioli, F.; Mando, M.A. The impact of sesquiterpene lactones and phenolics on sensory attributes: An investigation of a curly endive and escarole germplasm collection. *Food Chem.* **2016**, *199*, 238–245. [CrossRef]
10. Etesami, H.; Glick, B.R. Halotolerant plant growth–promoting bacteria: Prospects for alleviating salinity stress in plants. *Environ. Exp. Bot.* **2020**, *178*, 104124. [CrossRef]
11. Jiménez-Gómez, A.; García-Estévez, I.; García-Fraile, P.; Escribano-Bailón, M.T.; Rivas, R. Increase in phenolic compounds of *Coriandrum sativum* L. after the application of a *Bacillus halotolerans* biofertilizer. *J. Sci. Food Agric.* **2020**, *100*, 2742–2749. [CrossRef] [PubMed]
12. García-Fraile, P.; Carro, L.; Robledo, M.; Ramírez-Bahena, M.-H.; Flores-Félix, J.-D.; Fernández, M.T.; Mateos, P.F.; Rivas, R.; Igual, J.M.; Martínez-Molina, E.; et al. Rhizobium Promotes Non-Legumes Growth and Quality in Several Production Steps: Towards a Biofertilization of Edible Raw Vegetables Healthy for Humans. *PLoS ONE* **2012**, *7*, e38122. [CrossRef]
13. Ansari, R.A.; Rizvi, R.; Sumbul, A.; Mahmood, I. PGPR: Current Vogue in Sustainable Crop Production. In *Probiotics and Plant Health*; Springer International Publishing: Berlin/Heidelberg, Germany, 2017; pp. 455–472.
14. Jiménez-Gómez, A.; Flores-Félix, J.D.; García-Fraile, P.; Mateos, P.F.; Menéndez, E.; Velázquez, E.; Rivas, R. Probiotic activities of *Rhizobium laguerreae* on growth and quality of spinach. *Sci. Rep.* **2018**, *8*, 1–10. [CrossRef]
15. Zhang, N.; Yang, D.; Wang, D.; Miao, Y.; Shao, J.; Zhou, X. Whole transcriptomic analysis of the plant-beneficial rhizobacterium *Bacillus amyloliquefaciens* SQR9 during enhanced biofilm formation regulated by maize root exudates. *BMC Genom.* **2015**, *16*, 685. [CrossRef]
16. Flores-Félix, J.D.; Carro, L.; Cerda-Castillo, E.; Squartini, A.; Rivas, R.; Velázquez, E. Analysis of the Interaction between *Pisum sativum* L. and *Rhizobium laguerreae* Strains Nodulating This Legume in Northwest Spain. *Plants* **2020**, *9*, 1755. [CrossRef]
17. Ayuso-Calles, M.; García-Estévez, I.; Jiménez-Gómez, A.; Flores-Félix, J.D.; Escribano-Bailón, M.T.; Rivas, R. *Rhizobium laguerreae* improves productivity and phenolic compound content of lettuce (*Lactuca sativa* L.) under saline stress conditions. *Foods* **2020**, *9*, 1166. [CrossRef]
18. Overbeek, R.; Olson, R.; Pusch, G.D.; Olsen, G.J.; Davis, J.J.; Disz, T.; Edwards, R.A.; Gerdes, S.; Parrello, B.; Shukla, M.; et al. The SEED and the Rapid Annotation of microbial genomes using Subsystems Technology (RAST). *Nucleic Acids Res.* **2014**, *42*, D206–D214. [CrossRef]
19. Paterson, J.; Jahanshah, G.; Li, Y.; Wang, Q.; Mehnaz, S.; Gross, H. The contribution of genome mining strategies to the understanding of active principles of PGPR strains. *Fems Microbiol. Ecol.* **2017**, *93*. [CrossRef]
20. Cassán, F.; Vanderleyden, J.; Spaepen, S. Physiological and Agronomical Aspects of Phytohormone Production by Model Plant-Growth-Promoting Rhizobacteria (PGPR) Belonging to the Genus Azospirillum. *J. Plant Growth Regul.* **2013**, *33*, 440–459. [CrossRef]
21. Garcia-Fraile, P.; Seaman, J.C.; Karunakaran, R.; Edwards, A.; Poole, P.S.; Downie, J.A. Arabinose and protocatechuate catabolism genes are important for growth of *Rhizobium leguminosarum* biovar viciae in the pea rhizosphere. *Plant Soil* **2015**, *390*, 251–264. [CrossRef] [PubMed]

22. Granada, C.E.; Passaglia, L.M.P.; De Souza, E.M.; Sperotto, R.A. Is Phosphate Solubilization the Forgotten Child of Plant Growth-Promoting Rhizobacteria? *Front. Microbiol.* **2018**, *9*, 2054. [CrossRef]
23. Beneduzi, A.; Ambrosini, A.; Passaglia, L.M. Plant growth-promoting rhizobacteria (PGPR): Their potential as antagonists and biocontrol agents. *Genet. Mol. Biol.* **2012**, *35*, 1044–1051. [CrossRef] [PubMed]
24. Waśkiewicz, A.; Muzolf-Panek, M.; Goliński, P. Phenolic content changes in plants under salt stress. In *Ecophysiology and Responses of Plants Under salt Stress*; Ahmad, P., Azooz, M., Prasad, M., Eds.; Springer: New York, NY, USA, 2013; pp. 283–314.
25. Ambreetha, S.; Chinnadurai, C.; Marimuthu, P.; Balachandar, D. Plant-associated *Bacillus* modulates the expression of auxin-responsive genes of rice and modifies the root architecture. *Rhizosphere* **2018**, *5*, 57–66. [CrossRef]
26. Pérez-Montaño, F.; Del Cerro, P.; Jiménez-Guerrero, I.; López-Baena, F.J.; Cubo, M.T.; Hungria, M.; Megías, M.; Ollero, F.J. RNA-seq analysis of the *Rhizobium tropici* CIAT 899 transcriptome shows similarities in the activation patterns of symbiotic genes in the presence of apigenin and salt. *BMC Genom.* **2016**, *17*, 1–11. [CrossRef]
27. Mendis, H.C.; Thomas, V.P.; Schwientek, P.; Salamzade, R.; Chien, J.-T.; Waidyarathne, P.; Kloepper, J.; De La Fuente, L. Strain-specific quantification of root colonization by plant growth promoting rhizobacteria Bacillus firmus I-1582 and Bacillus amyloliquefaciens QST713 in non-sterile soil and field conditions. *PLoS ONE* **2018**, *13*, e0193119. [CrossRef]
28. Kumar, A.; Singh, J. Biofilms Forming Microbes: Diversity and Potential Application in Plant–Microbe Interaction and Plant Growth. In *Sustainable Development and Biodiversity*; Springer International Publishing: Berlin/Heidelberg, Germany, 2020; pp. 173–197.
29. Zhang, Y.; Ju, X.; Hong, T.; Li, Z. Nodule numbers, colonization rates, and carbon accumulations of white clover response of nitrogen and phosphorus starvation in the dual symbionts of rhizobial and arbuscular mycorrhizal fungi. *J. Plant Nutr.* **2019**, *42*, 2259–2268. [CrossRef]
30. Gamez, R.; Cardinale, M.; Montes, M.; Ramirez, S.; Schnell, S.; Rodriguez, F. Screening, plant growth promotion and root colonization pattern of two rhizobacteria (*Pseudomonas fluorescens* Ps006 and *Bacillus amyloliquefaciens* Bs006) on banana cv. Williams (Musa acuminata Colla). *Microbiol. Res.* **2019**, *220*, 12–20. [CrossRef]
31. Pereira, S.; Mucha, Â.; Gonçalves, B.; Bacelar, E.; Látr, A.; Ferreira, H.; Marques, G. Improvement of some growth and yield parameters of faba bean (*Vicia faba*) by inoculation with *Rhizobium laguerreae* and arbuscular mycorrhizal fungi. *Crop Pasture Sci.* **2019**, *70*, 595–605. [CrossRef]
32. El-Akhdar, I.; Elsakhawy, T.; Abo-Koura, H.A. Alleviation of Salt Stress on Wheat (Triticum aestivum L.) by Plant Growth Promoting Bacteria strains *Bacillus halotolerans* MSR-H4 and *Lelliottia amnigena* MSR-M49. *J. Adv. Microbiol.* **2020**, 44–58. [CrossRef]
33. Vidal, V.; Laurent, S.; Charles, F.; Sallanon, H. Fine monitoring of major phenolic compounds in lettuce and escarole leaves during storage. *J. Food Biochem.* **2018**, *43*, e12726. [CrossRef]
34. Mahmoudi, H.; Huang, J.; Gruber, M.Y.; Kaddour, R.; Lachaâl, M.; Ouerghi, Z.; Hannoufa, A. The Impact of Genotype and Salinity on Physiological Function, Secondary Metabolite Accumulation, and Antioxidative Responses in Lettuce. *J. Agric. Food Chem.* **2010**, *58*, 5122–5130. [CrossRef]

MDPI
St. Alban-Anlage 66
4052 Basel
Switzerland
Tel. +41 61 683 77 34
Fax +41 61 302 89 18
www.mdpi.com

Foods Editorial Office
E-mail: foods@mdpi.com
www.mdpi.com/journal/foods

www.ingramcontent.com/pod-product-compliance
Lightning Source LLC
LaVergne TN
LVHW070559100526
838202LV00012B/508